# 缅怀阿兰·布加特

阿兰·布加特（Alain BUGAT）是法国国家技术院名誉院长，也是法国原子能和替代能源委员会（CEA）的前任主席。作为核能的坚定捍卫者，布加特先生首倡善议，发起了本次中法三院在核能方面的合作。在报告的起草、中法三院专家的召集和报告的完善过程中，布加特先生投入了大量的时间，也花费了大量的精力。令我们深感悲痛的是，就在我们快马加鞭以至报告草创即就之际，布加特先生却于 2019 年 1 月溘然而逝。布加特先生的勃勃生气和超凡魅力，将永远留在我们的记忆中。

## In Memoriam of Alain BUGAT

Alain BUGAT, Honorary President of the National Academy of Technologies of France, Former General Administrator of the Commissariat à l'Energie Atomique (CEA) and a staunch defender of nuclear energy, initiated the three Academies collaboration on nuclear energy. He had paid much time and effort in shaping this report, bringing together experts from the three Academies and giving it credibility and excellence. To our great grief, he passed away in January 2019 at a time where the report was already at an advanced stage of completion. The memory of his energetic and charismatic personality rests in our minds.

# 核能与环境

## ——中法三院核能合作（第二期）

## Nuclear Energy and the Environment

### A Collaborative Programme between Chinese and French Academies（Phase Ⅱ）

中国工程院（Chinese Academy of Engineering）
法国国家技术院（National Academy of Technologies of France）编著
法国科学院（French Academy of Sciences）

科学出版社
北京

## 内 容 简 介

本书基于中法三院（中国工程院、法国国家技术院、法国科学院）对核能与环境领域的联合研究成果，整理和分析了大量数据，着重研究了从铀矿开采到放射性废物处理的整个过程中核能对环境的影响，包括正常和事故工况，以及核废物的环境影响，并对这些影响进行了全面分析。中法三院就核能与环境领域相关问题提出了见解和建议，将有效促进核能的健康发展和进一步提升中法两国对世界核能技术发展的贡献。

本书主要面向关注核能发展的国家的科研工作者、管理人员和有志投身核能事业的青年学子，力图立足科学立场阐释公众关心的问题，对于提高国际核能领域科学技术界的交流水平、增强在核能和平利用上的共识具有重要意义。

**图书在版编目(CIP)数据**

核能与环境：中法三院核能合作：第二期 / 中国工程院，法国国家技术院，法国科学院编著. —北京：科学出版社，2023.9

ISBN 978-7-03-076481-2

Ⅰ.①核… Ⅱ.①中… ②法… ③法… Ⅲ.①核污染-环境污染-相关分析-国际科技合作-中、法 Ⅳ.①X591

中国国家版本馆 CIP 数据核字（2023）第 185323 号

责任编辑：周 涵 / 责任校对：彭珍珍
责任印制：吴兆东 / 封面设计：无极书装

科 学 出 版 社 出版
北京东黄城根北街 16 号
邮政编码：100717
http://www.sciencep.com
北京中科印刷有限公司印刷
科学出版社发行 各地新华书店经销
*
2023 年 9 月第 一 版 开本：787×1092 16
2024 年 1 月第二次印刷 印张：17 1/2
字数：309 000

**定价：178.00 元**

（如有印装质量问题，我社负责调换）

## About the book

The book is based on a collaborative study on *Nuclear Energy and the Environment* that was conducted by the Chinese Academy of Engineering, the National Academy of Technologies of France and the French Academy of Sciences. The authors, by acquiring and analyzing adequate data, studies impacts of nuclear energy on the environment during the full life cycle of nuclear power from uranium mining to radioactive waste disposal, and provides a comprehensive evaluation of the impacts of nuclear power on the environment, including the impacts in normal or accidental situations the impact of radioactive waste. Insights and recommendations concerning nuclear power and the environment provided by the three Academies shall be beneficial for healthy development of nuclear power, and will add to contributions of China and France to the global nuclear power.

The book is mainly for researchers and managers who are engaging in nuclear power, and young students who are to dedicate themselves to the area. The book pursues to address the public concerns about nuclear energy from a scientific perspective. It is of important significance to facilitate academic exchanges in the community of international nuclear science and technology and to enhance consensus on peaceful uses of nuclear energy.

# 作 者 简 介

作者介绍可在法国国家技术院、法国科学院、中国工程院的网站上找到。以下只列出作者的基本任职信息。

法国国家技术院

http://www. academie-technologies. fr/en/members

**Alain BUGAT**（项目联合组长）法国国家技术院荣誉院长、院士。曾任法国原子能署主席、NUCADVISOR 咨询公司的联合创始人兼总裁。

**Yves BAMBERGER** 法国国家技术院院士，法国电力公司研发部前主任。

**Pascal COLOMBANI** 法国国家技术院院士。A. T Kerney（巴黎）高级顾问，TechnipFMC 董事会成员，曾任 CEA 行政主管、阿海珐集团监督委员会主席。

**Bernard ESTEVE** 法国国家技术院院士，道达尔公司前核工业顾问，目前任 B. E. 咨询公司董事长。

**Gerard GRUNBLATT** 法国国家技术院院士，阿尔斯通公司超导应用前经理。

**Patrick LEDERMANN** 法国国家技术院院士，阿尔斯通印度有限公司前董事总经理。

**Philippe PRADEL** 法国国家技术院院士，法国 Engie 集团副总裁。

**Bruno REVELLIN-FALCOZ**（国际协调员）法国国家技术院荣誉院长、院士，达索航空公司前副总裁。

**Bernard TARDIEU** 法国国家技术院院士，科因贝利公司荣誉总裁。

**Dominique VIGNON** 法国国家技术院院士，法玛通公司前总裁兼总经理。国际原子能机构国际安全咨询小组前成员、法国核能学会前任主席。

**法国科学院**

http：//www. academie-sciences. fr/en/Members/members-of-the-academie-des-sciences. html

**Sébastien CANDEL**（项目联合组长）法国科学院院长，工程科学专家，中央理工－高等电力学院（萨克雷大学）荣誉教授，近期新任法国电力公司科学委员会主席。

**Edouard BREZIN** 法国科学院荣誉院长、院士，统计物理和粒子物理专家，巴黎高等师范学校荣誉教授。

**Robert GUILLAUMONT** 法国科学院院士，放射化学专家，奥赛大学荣誉教授，国家评估委员会成员。

**中国工程院**

http://en. cae. cn/en/Member/Member/

**赵宪庚**（项目联合组长）中国工程院院士，中国工程院原副院长。

**叶奇蓁**（项目副组长）中国工程院院士，核反应堆、核电技术专家，曾任秦山核电站一期总工程师。

**中国工作组**

中国工作组成员来自中国核工业集团有限公司（CNNC）、生态环境部核与辐射安全中心（NRSC）、中国核电工程有限公司（CNPE）、中国原子能科学研究院（CIAE）。

组长：雷增光（中国核工业集团有限公司）

总执笔人：刘森林（中国原子能科学研究院）

第 1 章执笔人：姜子英（中国原子能科学研究院）

第 2 章执笔人：陈晓秋（生态环境部核与辐射安全中心），杨端节（生态环境部核与辐射安全中心），张燕齐（中国原子能科学研究院）

第 3 章执笔人：刘新华（生态环境部核与辐射安全中心），魏方欣（生态环境部核与辐射安全中心），张振涛（中国原子能科学研究院）

第 4 章执笔人：陈巧艳（中国核电工程有限公司），薛娜（中国核电工程有限公司），喻新利（中国核电工程有限公司），王辉（中国核电工程有限公司）

第 5 章执笔人：柴国旱（生态环境部核与辐射安全中心），李静晶（中国原子能科学研究院）

**技术与支持组**

法国国家技术院：Wolf GEHRISCH（技术秘书）

法国科学院：Jean-Yves CHAPRON（技术秘书）

中国工程院：王振海，田琦，宗玉生，刘玮，谢光辉，张宁，王浩闻，周亚琳，徐琳，李艳杰，常润华

中国工程物理研究院：彭现科，郑刚阳

中国核工业集团有限公司：钱天林，刘仲华，周觅

中国原子能科学研究院：喻宏，尹忠红，张徐璞，夏芸，夏梦蝶（技术秘书），王新燕

**致谢**

全体作者对以下人士仔细阅读报告初版并提供了诸多有用的建议表示感谢：Yves BRECHET（法国科学院），Antoine DANCHIN（法国科学院），Jean-Claude DUPLESSY（法国科学院），Marc FONTECAVE（法国科学院），Guy LAVAL（法国科学院），Olivier PIRONNEAU（法国科学院），Jean SALENÇON（法国科学院），Jean FRÊNE（法国国家技术院），评审委员会主席和全体成员，Pierre LAMICQ（法国国家技术院），Bernard PAYEN（法国国家技术院），Pierre TOULHOAT（法国国家技术院）。

全体作者也对以下人士所提供的宝贵意见和建议表示感谢：杜祥琬（中国工程院），潘自强（中国工程院），徐銤（中国工程院），陈念念（中国工程院），李冠兴（中国工程院），于俊崇（中国工程院），侯立安（中国工程院），钮新强（中国工程院），邓建军（中国工程院），张华祝（中国核能行业协会），程慧平（中国核工业集团有限公司），张金涛（中国核工业集团有限公司），毛亚蔚（中国核电工程有限公司），霍小东（中国核电工程有限公司），任晓娜（中国辐射防护研究院），王驹（核工业北京地质研究院），刘健（核工业北京地质研究院），周连全（中国原子能科学研究院），杨勇（中国原子能科学研究院），任逸（中国原子能科学研究院），陈延鑫（核工业研究生部）。

# About the authors

Information concerning the authors curriculum vitae may be obtained from the web sites of National Academy of Technologies of France, French Academy of Sciences and Chinese Academy of Engineering. Short biographical data are given below.

National Academy of Technologies of France (AT)

http://www. academie-technologies. fr/en/members

**Alain BUGAT** (Study co-leader) was Honorary President of the National Academy of Technologies of France. He was Former Head of the Commissariat à l'Energie Atomique (CEA) and Co-founder and President of NUCADVISOR.

**Yves BAMBERGER** is Member of the National Academy of Technologies of France and Former Director of Research and Development at EDF.

**Pascal COLOMBANI** is Member of the National Academy of Technologies of France. He is Senior Advisor of A. T. Kearney Paris, and Member of the Board of TechnipFMC, Former Head of the Commissariat à l'Energie Atomique (CEA), Former Chairman of the Supervisory Board of AREVA.

**Bernard ESTEVE** is Member of the National Academy of Technologies of France and Former Nuclear Counsellor for Total. He is currently President of B. E. Consult.

**Gerard GRUNBLATT** is Member of the National Academy of Technologies of France. He is Former Head of Superconductivity Applications at ALSTOM.

**Patrick LEDERMANN** is Member of the National Academy of Technologies of France and Former Managing Director of ALSTOM India limited.

**Philippe PRADEL** is Member of the National Academy of Technologies of France and Vice-President of ENGIE Nucléaire, France.

**Bruno REVELLIN-FALCOZ** (International coordinator) is Member and Honorary President of the National Academy of Technologies of France. He is Former Vice-President and Director-General of Dassault Aviation.

**Bernard TARDIEU** is Member of the National Academy of Technologies of France and Honorary President of COYNE and BELLIER.

**Dominique VIGNON** is Member of the National Academy of Technologies of France and of the World Nuclear Academy, Former Chairman and Chief Executive Officer of Framatome, Former Member of the International Safety Advisory Group to the IAEA and Past President of the French Nuclear Energy Society.

**French Academy of Sciences (AS)**

http://www.academie-sciences.fr/en/Members/members-of-the-academie-des-sciences.html

**Sébastien CANDEL** (Study co-leader) is President of the French Academy of Sciences. He is a specialist in Engineering Sciences, University Professor Emeritus at Centrale Supélec, University Paris-Saclay. He has recently been appointed Chairman of the Scientific Council of EDF.

**Edouard BREZIN** is Member and Past President of the French Academy of Sciences. He is a specialist in Statistical and Particle Physics and Professor Emeritus at Ecole Normale Supérieure.

**Robert GUILLAUMONT** is Member of the French Academy of Sciences. He is a specialist in Radiochemistry, Honorary Professor at the University of Orsay and Member of the Commission Nationale d'Evaluation.

**Chinese Academy of Engineering (CAE)**

http://en.cae.cn/en/Member/Member/

**ZHAO Xiangeng** (Study co-leader) is Member and Past Vice President of CAE.

**YE Qizhen** (Assistant co-leader) is Member of CAE. He is a specialist in the field of nuclear reactor and nuclear power generation technology. He was Chief Design

Engineer of the Qinshan Nuclear Power Project.

### Chinese Working Team

Members of this team work for CNNC (China National Nuclear Corporation), NRSC (Nuclear and Radiation Safety Center), CNPE (China Nuclear Power Engineering Corporation), and CIAE (China Institute of Atomic Energy) .

LEI Zhengguang (CNNC) (Team leader), LIU Senlin (CIAE) (General editor)

JIANG Ziying (CIAE) (PIC of Chapter 1); CHEN Xiaoqiu (NSC) (PIC of Chapter 2), YANG Duanjie (NSC), ZHANG Yanqi (CIAE); LIU Xinhua (NSC) (PIC of Chapter 3), WEI Fangxin (NSC), ZHANG Zhentao (CIAE); CHEN Qiaoyan (CNPE) (PIC of Chapter 4), XUE Na (CNPE), YU Xinli (CNPE), WANG Hui (CNPE); CHAI Guohan (NSC) (PIC of Chapter 5), LI Jingjing (CIAE)

### Technical and Support Team

Wolf GEHRISCH (AT) (Technical secretary), Jean-Yves CHAPRON (AS) (Technical secretary), WANG Zhenhai (CAE), TIAN Qi (CAE), ZONG Yusheng (CAE), LIU Wei (CAE), XIE Guanghui (CAE), ZHANG Ning (CAE), WANG Haowen (CAE), ZHOU Yalin (CAE), XU Lin (CAE), LI Yanjie (CAE), CHANG Runhua (CAE), PENG Xianke (CAEP), ZHENG Gangyang (CAEP), QIAN Tianlin (CNNC), LIU Zhonghua (CNNC), ZHOU Mi (CNNC), YU Hong (CIAE), YIN Zhonghong (CIAE), ZHANG Xupu (CIAE), XIA Yun (CIAE), XIA Mengdie (CIAE) (Technical secretary), WANG Xinyan (CIAE)

### Acknowledgments

The authors wish to thank Yves BRECHET (AS), Antoine DANCHIN (AS), Jean-Claude DUPLESSY (AS), Marc FONTECAVE (AS), Guy LAVAL (AS), Olivier PIRONNEAU (AS), Jean SALENÇON (AS); Jean FRÊNE (AT), President of the Reviewing Committee and all members of this Committee, Pierre LAMICQ (AT), Bernard PAYEN (AT), Pierre TOULHOAT (AT) for their careful reading of the initial version of this report and for their many helpful comments.

The authors would also like to express their gratitude to DU Xiangwan (CAE), PAN Ziqiang (CAE), XU Mi (CAE), CHEN Niannian (CAE), LI Guanxing (CAE), YU Junchong (CAE), Hou Lian (CAE), NIU Xinqiang (CAE), DENG Jianjun (CAE), ZHANG Huazhu (CNEA), CHENG Huiping (CNNC), ZHANG Jintao (CNNC), MAO Yawei (CNPE), HUO Xiaodong (CNPE), REN Xiaona (CIRP), WANG Ju (BRIUG), LIU Jian (BRIUG), ZHOU Lianquan (CIAE), YANG Yong (CIAE), REN Yi (CIAE), CHEN Yanxin (CIAE) for providing valuable comments and remarks.

# 前　　言

本报告由中法三院联合发布，主题是核能与环境。2017 年 8 月，中法三院发布了第一期《关于核能未来的联合建议》，讨论了核能的诸多问题，并重点从技术方面就核能未来的发展方向提出了联合建议。本报告是中法三院第二次关于核能问题的联合研究成果，报告力求客观地阐述核能的诸多科学和技术问题（包括核能在未来能源结构中的地位、核能的效益、核能的优势和劣势、核能的研发以及核能的技术、安全和工程等）以及社会问题（包括教育、培训、风险感知、公众认可等）。尽管如此，本报告也不可能面面俱到。

尽管环境问题对核工业的未来发展至关重要，然而第一期报告并未充分考虑环境问题。有鉴于此，中法三院认为有必要继续开展合作，进一步对环境这一重要问题进行研究。本次中法双方联合研究的重点是讨论正常和事故工况下核能（包括核废物）对环境的影响，并对这些影响进行全面分析，其中这些影响在中法两国具有相似的性质。作为未来发展的一个重要因素，能源生产的经济性取决于地域条件，而中法两国的地域条件截然不同，因此本次联合研究决定暂不讨论这个问题。

本报告由中法联合研究小组以中英法三语同时编写，现将中英双语版本正式出版。

# Foreword

This report on nuclear energy and the environment is jointly released by the Chinese Academy of Engineering, the National Academy of Technologies of France and the French Academy of Sciences. It is a second collaborative study of the three Academies on nuclear energy issues. The first report *Joint Recommendations for the Nuclear Energy Future* issued in August 2017, covered many aspects of nuclear energy and offered joint, mostly technical recommendations for the direction of nuclear energy in the future. It was an attempt to provide an objective overview of many scientific and technological issues on nuclear energy (its position in the future energy mix, benefits, strengths and weak points, research and development perspectives, technology and safety, engineering etc.), as well as societal issues (education, training, risk perception, public awareness etc.). But it was admittedly far from being exhaustive.

Environmental issues were not considered in sufficient depth in the previous report despite their being crucial for the future of this industry, and the Academies felt it necessary to pursue their cooperation and further address these important issues. They decided to focus the joint effort on the environmental impacts of nuclear energy in normal and accidental situations, including waste, and provide a comprehensive analysis of these issues which are essentially similar in France and China. However, the economics of energy production, which also constitutes an important factor for the future, is determined by local and regional conditions, which are fairly different between these two countries, and it was deliberately decided not to address this issue in this joint study.

# 目　　录


综述与建议 …… 1

0.1　核电厂和燃料循环设施正常运行期间的环境影响 …… 1

0.1.1　核能对全球环境的影响 …… 1

0.1.2　核能对当地环境的影响 …… 2

0.2　核电厂及核燃料循环设施在事故情况下的影响 …… 3

0.3　放射性废物管理的影响 …… 4

0.4　利用核与辐射安全、安保措施来降低环境影响 …… 5

第1章　引言 …… 6

1.1　能源需求趋势 …… 6

1.2　脱碳承诺，以及不同发电能源 $CO_2$ 排放情况 …… 8

1.3　环境保护要求核能可持续发展 …… 8

1.4　本报告内容和结构 …… 10

第2章　核电厂和核燃料循环设施正常运行期间的环境影响 …… 11

2.1　如何衡量核能的环境影响 …… 12

2.1.1　核能对环境的主要影响 …… 12

2.1.2　环境释放 …… 13

2.1.3　评价 …… 13

2.1.4　方法 …… 14

2.2　核能的流出物、辐射影响及解决方案 …… 14

2.2.1　核电厂的流出物及辐射影响 …… 14

2.2.2　核燃料循环的流出物及辐射影响 …… 17

2.2.3　辐射监测和监控 …… 19

2.2.4　电离辐射的生物效应 …… 20

2.2.5 放射性物质的运输…… 20
2.2.6 利益相关方参与…… 21
2.3 核能发电与其他能源发电的环境影响比较…… 22
2.3.1 土地占用…… 23
2.3.2 建造材料…… 24
2.3.3 取水和用水…… 25
2.3.4 常规退役废物…… 28
2.3.5 关键材料…… 28
2.4 新技术展望…… 28
2.4.1 减少燃烧燃料的碳排放…… 29
2.4.2 嬗变技术…… 30
2.4.3 其他先进技术…… 32
2.5 结论…… 32
附录 2-1 关于中国的具体情况 …… 34
附录 2-2 核厂址周围大型流行病学调查（案例与结果） …… 36
**第 3 章 乏燃料和放射性废物管理 …… 38**
3.1 放射性废物管理的原则、策略和框架…… 39
3.1.1 放射性废物管理原则…… 39
3.1.2 放射性废物管理策略…… 40
3.1.3 放射性废物管理框架…… 42
3.2 放射性废物的特性和分类…… 44
3.2.1 乏燃料和/或后处理放射性废物 …… 44
3.2.2 放射性废物的特性…… 45
3.2.3 放射性废物的分类…… 47
3.3 放射性废物处理和流出物排放…… 48
3.3.1 放射性废物最小化…… 48
3.3.2 流出物排放…… 49
3.4 放射性废物处置…… 50
3.4.1 极低放废物（VLLW） …… 50

3.4.2　短寿命低中放废物（LIL-SLW）　……………………………… 51
3.4.3　长寿命低放废物（LL-LLW）……………………………………… 52
3.4.4　长寿命中放废物（IL-LLW）和高放废物（HLW）　………… 52
3.4.5　铀燃料循环（铀矿开采）前端产生的仅含有天然放射性核素的放射性废物……………………………………………… 55
3.5　开式/闭式核燃料循环产生的废物 ……………………………… 56
3.6　新技术……………………………………………………………… 57
3.7　结论………………………………………………………………… 58
附录 3　放射性废物处置设施 ………………………………………… 59
A3.1　法国 ……………………………………………………………… 59
A3.2　中国 ……………………………………………………………… 60
**第 4 章　严重核事故 ………………………………………………………… 62**
4.1　核电厂严重事故…………………………………………………… 63
4.1.1　三里岛核电厂事故……………………………………………… 63
4.1.2　切尔诺贝利核电厂事故………………………………………… 64
4.1.3　福岛第一核电厂事故…………………………………………… 66
4.2　为降低事故环境影响采取的改进措施…………………………… 68
4.2.1　反应堆技术提升………………………………………………… 68
4.2.2　核电厂在福岛第一核电厂事故后的行动……………………… 69
4.2.3　严重事故管理…………………………………………………… 70
4.2.4　对于未来类似严重事故的见解………………………………… 72
4.3　结论………………………………………………………………… 73
附录 4－1　第三代压水堆严重事故专用预防和缓解措施 ………… 74
附录 4－2　中国内陆核电厂的安全性 ……………………………… 75
附录 4－3　中国严重事故后应急组织管理 ………………………… 76
A4－3.1　中国的三级核应急体系 …………………………………… 77
A4－3.2　核应急监测体系 …………………………………………… 77
**第 5 章　核安全与环境 ……………………………………………………… 79**
5.1　核电厂的安全性及其环境影响…………………………………… 80

5.1.1 严重事故及其后果…… 81
5.1.2 风险指引的纵深防御体系 …… 82
5.1.3 新的安全威胁因素…… 84
5.2 核电厂选址…… 85
5.3 安全责任及政府职责…… 86
5.3.1 运营单位的主要责任…… 86
5.3.2 政府和监管机构的职责和作用…… 87
5.4 核安全与公众接受度…… 87
5.5 结论…… 88
附录 5-1 核安全监管体系 …… 89
附录 5-2 中国采取的行动 …… 90
**第 6 章 结论** …… **92**
**参考文献** …… **96**
**词汇表** …… **101**
**后记** …… **106**

# Contents

**Synthesis and recommendations** ······ **108**

0.1 Impacts from NPPs and nuclear fuel cycle facilities under normal operation ······ 109

0.1.1 Nuclear energy and global impacts on the environment ······ 109

0.1.2 Nuclear energy and local impacts on the environment ······ 109

0.2 Impacts from NPPs and nuclear fuel cycle facilities in accidental situations ······ 111

0.3 Impacts from radwaste management ······ 112

0.4 Nuclear and radiation safety/security as a tool to prevent impacts on the environment ······ 113

**Chapter 1 Introduction** ······ **115**

1.1 Trends in energy demand ······ 115

1.2 Decarbonization commitment and $CO_2$ emissions by various energy options ······ 117

1.3 Environmental protection as a requirement to make nuclear energy sustainable ······ 119

1.4 Report contents and organization ······ 120

**Chapter 2 Environmental impacts during normal operations of nuclear power plants and nuclear fuel cycle facilities** ······ **122**

2.1 How to measure impacts of nuclear energy on the environment ······ 123

2.1.1 Main impacts of nuclear energy to the environment ······ 123

2.1.2 Releases to the environment ······ 124

2.1.3 Assessments ······ 125

2.1.4 Methodology ······ 126

2.2 Effluents, radiological impacts of nuclear energy and solutions ········ 126
2.2.1 Effluents and radiological impacts of NPPs ························ 126
2.2.2 Effluents and radiological impacts of nuclear fuel cycle ·········· 129
2.2.3 Radiation monitoring and surveillance ······························ 131
2.2.4 Biological effect of ionizing radiation ································ 132
2.2.5 Transportation of radioactive materials ······························ 133
2.2.6 Participation of stakeholders ········································· 134
2.3 Environmental impacts of nuclear energy compared to other sources of electricity ······························································ 135
2.3.1 Land occupation ························································ 137
2.3.2 Materials used for construction ······································ 138
2.3.3 Water withdrawal and consumption ································ 139
2.3.4 Conventional waste from decommissioning ······················· 143
2.3.5 Critical materials ························································ 144
2.4 New technology perspectives ············································· 144
2.4.1 Reducing Carbon emissions of burnt fuels ························ 144
2.4.2 Transmutation technologies ·········································· 145
2.4.3 Other advanced technologies ········································ 148
2.5 Conclusions ································································· 149
Appendix 2-1 More specific considerations about the Chinese situation ······ 151
Appendix 2-2 Large scale epidemiological studies around nuclear sites (cases and results) ················································ 153
**Chapter 3 Spent fuel and radwaste management** ······························ **157**
3.1 Principles, strategies and framework of radwaste management to prevent environmental impacts ·········································· 159
3.1.1 Principles of radwaste management ································ 159
3.1.2 Strategies of radwaste management ································ 159
3.1.3 Frameworks of radwaste management with respect to the environment ·············································· 163
3.2 Specific characteristics and classification of radwaste ························ 166

3.2.1 Spent fuel or reprocessing radwaste …… 166
3.2.2 Specific characteristics of radwaste versus the environment …… 168
3.2.3 Classification of radwaste versus the environment …… 170
3.3 Processing and discharge of radwaste …… 172
3.3.1 Minimization of radwaste …… 172
3.3.2 Discharge of effluents …… 174
3.4 Disposal of radwaste …… 174
3.4.1 Very low-level waste (VLLW) …… 174
3.4.2 Low and intermediate level-short lived waste (LIL-SLW) …… 175
3.4.3 Low level-long lived waste (LL-LLW) …… 177
3.4.4 Intermediate level-long lived waste (IL-LLW) and high level waste (HLW) …… 178
3.4.5 Radioactive waste containing only natural radionuclides from the front-end of uranium fuel cycle (uranium mining) …… 181
3.5 Open/closed nuclear fuel cycle …… 184
3.6 New technologies …… 185
3.7 Conclusions …… 186
Appendix 3 …… 188
A3.1 French side …… 188
A3.2 Chinese side …… 189
**Chapter 4 Severe nuclear accidents …… 191**
4.1 Severe accidents …… 192
4.1.1 Three Mile Island accident …… 193
4.1.2 Chernobyl accident …… 195
4.1.3 Fukushima Daiichi accident …… 197
4.2 Improvements to make nuclear energy free of environmental impacts in case of accident …… 200
4.2.1 Improvements in reactor technologies …… 200
4.2.2 NPP action after Fukushima Daiichi accident …… 202
4.2.3 Severe accident management …… 203

4.2.4 Insights on similar severe accidents in future …… 206
4.3 Conclusions …… 207
Appendix 4-1 Dedicated prevention and mitigation measures for severe accidents of Gen-Ⅲ NPPs …… 208
Appendix 4-2 Safety of inland NPP in China …… 210
Appendix 4-3 Emergency management after severe accidents in China …… 212
A4-3.1 Three level nuclear emergency system in China …… 213
A4-3.2 Nuclear emergency monitoring system …… 214
**Chapter 5 Nuclear safety and the environment** …… **216**
5.1 The safety of nuclear power plants and their environmental impact …… 218
5.1.1 Severe accidents and their external consequences …… 218
5.1.2 Risk-informed defence in depth …… 221
5.1.3 New safety threats …… 223
5.2 NPPs sitings …… 224
5.3 Responsibility for safety and role of the government …… 226
5.3.1 The prime responsibility of the operator …… 226
5.3.2 The role of the government and the regulator …… 228
5.4 Nuclear safety, and public understanding …… 229
5.5 Conclusions …… 230
Appendix 5-1 Nuclear safety regulation system …… 231
Appendix 5-2 Actions taken in China …… 232
**Chapter 6 Conclusions** …… **235**
**References** …… **240**
**Glossary** …… **245**
**Postscript** …… **250**

# 综述与建议

2017 年 8 月，中法三院的第一期报告就核能的未来提出了一系列联合建议。2017 年 9 月，第一期报告在维也纳举行的国际原子能机构大会期间进行了发布，发布活动也成为本次大会的一项“场外边会”。本报告是中法三院核能合作的第二期报告，具体地讨论了核能循环对环境的影响，以回应社会对将核能的环境影响纳入所有人类活动的强烈期待。关于核能的环境影响，公众对核电厂（nuclear power plant，NPP）在正常运行或事故工况下的放射性影响以及长期时间范围内地质处置后放射性废物中的放射性元素重回生物圈等问题，表示出种种担忧。

从铀矿开采到放射性废物处理，中法三院对核能的整个过程进行了考察，评估了正常或事故工况下核能给全球和当地环境带来的短期和长期影响，并考虑了核能发展对人类和生态系统带来的多种后果。本报告总结了相关经验教训，以及为可持续的保护环境而已经采取或可能采取的行动。在此，我们的结论是，第三代核电厂及其相关设施，以及未来的第四代核电厂将有可能减少环境影响。

思考和建议总结如下。中法三院认为，报告提出的建议大多与核能利益攸关方所采取的行动相对应。中法三院指出，核工业界应当实施报告提出的这些有价值的行动，并在某些情况下强化这些行动。

## 0.1 核电厂和燃料循环设施正常运行期间的环境影响

### 0.1.1 核能对全球环境的影响

根据几项生命周期分析（life cycle analyses，LCA），核能每兆瓦产生的二氧化碳较少，与水力发电一样少，明显优于光伏（photovoltaics，PV）发电，仅略高于风能。但是，必须指出，间歇性能源在无法获得时需要得到补偿，这就改变了其环境绩效。就建筑标准材料和关键金属材料的消耗而言，同样的能源生产，核能所需的量比光伏和风能要少得多。放射性影响主要来自其向环境释放的放射性气体（稀有气体、氚、氡等）和液体排放物（主要是氚）。辐射对公众的影响只占天然辐射源总体影响的很小一部分，不到1%。关于低剂量和极低剂量/剂量率照射的长期影响存在着争论，然而，根据世界上大多数流行病学方面的研究认为没有证据表明其会对生物产生影响。分子流行病学方面的研究考虑了细胞水平层面的确定影响，这方面的研究可能更有效，应予以鼓励支持。同样值得注意的是，有些物种，如昆虫，可以承受很高水平的辐射。

### 0.1.2 核能对当地环境的影响

化石燃料发电厂（特别是使用煤炭或褐煤的电厂）排放大量的空气污染物，如颗粒物、氮氧化物、硫氧化物、重金属和各类其他化学物质，但核电厂并非如此。在这方面，燃煤电厂主要以气态氡的形式释放大量的天然放射性，而且氡的固体废物中含有大量的铀和钍，需要作为放射性废物（这里指人为活动引起天然放射性水平增加的工业废物（technologically enhanced naturally occurring radioactive material，TENORM））加以管理。因此，如果核能导致化石燃料电厂关闭，这实际上对当地会产生积极的影响。没有了排放物可以显著地改善空气质量，并减少酸雨等对环境的破坏。核能对环境的影响主要与前端设施有关，其中核燃料循环的前端是指铀矿的开采一直到浓缩铀交付给核燃料组件生产商（氧化铀（uranium oxide，$UO_x$）燃料，其中包括为制造铀钚混合氧化物（mixed uranium-plutonium oxide，MOX）燃料而开展的钚的处理）的阶段。

正如第2章定量分析的那样，建造反应堆产生的非放射性技术废物低于建造风力涡轮机或光伏装置产生的非放射性技术废物。

与光伏或风力发电场相比，核能发电占用的土地也大大减少，大约2/3的用地是由于采矿和退役核电厂。

从河流中抽取电站冷却水的问题需要特别注意，核电厂生产每兆瓦时（MWh）电力所需的冷却水要高于化石燃料电站。考虑到水资源的可用性，在内陆核电厂选址时需要考虑水资源压力和水温的升高。一般而言，大部分的冷却水都流回到了河里，但目前已经出现了气候变暖和天气干燥的现象，这会迫使核电厂偶尔需要在低于标称功率的条件下运行。因此，应当对全球变暖的潜在影响进行谨慎预测。

中法三院认为，正确评估核能对环境的影响需要：

● 在所有情况下，将核活动引起的照射与天然照射进行比较。

● 在新核设施投入运行之前，应进行本底流行病学研究，这对于任何事故后流行病学研究的比较分析、辐射风险分析和回应公众的关切是至关重要和必要的。

● 内陆核电厂选址应考虑水资源压力和未来气候变化。

此外，中法三院还提出了一个一般性建议，希望减少核能的环境影响：

● 积极发展先进核技术，可以减少核燃料循环运行的前端对环境的影响。由于核燃料循环前端活动的影响大于后端活动（即乏燃料的管理直至地质处置）的影响，基于低铀耗的第四代核电投入使用后将有利于改善环境。快中子反应堆或多次循环乏燃料有可能大大减少这些影响。目前，需要进行先进核技术商业化有关的筹备工作。

一般而言，除了用水，核能装机每兆瓦使用的材料有限。在正常运行和整个燃料循环过程中，核能对环境的辐射和非辐射影响也是有限的。

## 0.2　核电厂及核燃料循环设施在事故情况下的影响

核能对环境的主要影响是核能发展历史上标志性的严重事故（在国际核事件分级表（International Nuclear Event Scale，INES）中为 6 级或 7 级）造成的。切尔诺贝利核电厂和福岛第一核电厂事故对公众舆论和全球核能发展产生了巨大影响。我们从这些事故和三里岛事故（INES 上 5 级）中吸取了相关教训并在后期的反应堆设计和操作程序上进行了重大技术变革。这些技术变革在第三代核电厂中得以实施，并在合理范围内对运行中的反应堆进行了改造。虽然未来可能发生严重事故，但环境风险已大大降低，只局限于核电厂的范围内。

中法三院建议：

● 继续研究导致严重事故的机制（内部事件，如（功率）极限增长、冷却剂丧失，或外部事件，如地震、飞机失事、恐怖袭击等），并为预防和减轻事故提供支持。

目前，应该开展进一步有关保持压力容器的完整性的研究，或开发耐事故燃料（accident tolerant fuel，ATF）。

● 实施严重事故管理指南，积累更多经验，实施预防和缓解措施，以应对核电厂和多机组事故中的大规模失效问题，增强应急响应能力。

## 0.3　放射性废物管理的影响

核能会产生短期和长期的放射性废物。管理前者可以利用工业方式，在近地表贮存库中处置；后者的管理取决于其放射活性。最高放射性废物（乏燃料或后处理产生的中高放废物）计划在深层地层中处置。开采和提炼铀产生的放射性废料得到了妥善处理（主要是采矿过程中产生的废料，采取就地处理的方式）。核能对环境的直接影响主要来自处理和包装未经处理的废物所排放的污水。按照目前的操作，这些活动对当地和全球的健康和环境影响非常小。大型数据库支持的许多模拟结果表明，如果有长期的延迟影响，预计会小于自然辐射的影响。然而与公众认为的一样，管理放射性废物是核能的主要挑战之一。

为加深放射性废物对环境实际影响的认识，中法三院建议：

● 考虑到核燃料循环前端和后端产生的废物以及时间尺度，应改进评估所有环境影响（放射学和化学）及其相关风险的方法。

为支持这项一般性建议，中法三院建议：

● 为更好地处理环境问题，应定义量化参数来描述与放射性废物有关的危害。

● 进一步改进研发（research and development，R&D）方案，更好地了解核能对生态系统的辐射和化学影响（生态系统构成要素的可逆性、可恢复性、生物可用性……）。

● 应采用全面和负责任的制度来保护环境（包括立法、主管和独立机构、筹资程序），让公众可以清楚地获取上述信息。

一般而言，只有应用最好的现有技术（best available technology，BAT，这技术可靠且技术成熟度高），才能在每一个环节中限制放射性核素的释放。

## 0.4 利用核与辐射安全、安保措施来降低环境影响

核安全的一个主要目标，是消除严重事故条件下放射性物质大量释放到环境的可能性，这是核能面临的一个主要问题。作为设计者、运行者和安全监管部门的责任，核与辐射安全在环境保护中扮演着关键的角色。作为政府的主要职责，核安保的目标是防止针对核设施的恶意行动，其中这种恶意行动可能会导致放射性物质的释放。

中法三院建议，核设施的所有者：

- 测试现有核设施对外部事件的恢复能力，这一能力应高于设计基准中考虑的水平；
- 升级现有核设施，达到新设施设定的安全目标，且合理可行；
- 对所有设施实施基于风险的纵深防御，包括超出设计基准的防御；
- 对核安全管理体系进行额外的外部审查，不完全依赖安全部门的审查。

由于环境保护是一个重大敏感问题，建议核管理机构：

- 组建透明的核安全监管机构，并保证沟通的透明化；
- 与当地政府和公众建立长期对话机制。

中法三院认为，各方应共同努力，向公众宣传核能，开展核能教育，特别是与环境影响有关的问题。

满足以上条件后，中法三院认为，要想保护环境，最好使用混合能源，包括核能和可再生能源。

# 第1章　引　　言

核能具有很多优势，特别是在作为可调峰电源时，其温室气体排放水平和大气污染物排放量极低。当前，全球气候变化可能是人类面对的最重要的环境问题，而这恰恰是核能最显著的优势。考虑到世界很多地区的空气质量在持续恶化，核能的大气污染物排放水平极低这一优势就显得格外重要。但在另一方面，同所有其他能源一样，核燃料循环也会产生环境影响。本报告对这类影响进行了综合评估，并讨论了如何对其进行限制和控制。因此，本报告专门关注环境和安全议题，而未能对其他问题面面俱到。

本章分为 3 节。1.1 节简要论述了能源需求趋势。1.2 节讨论了脱碳承诺和各类能源的 $CO_2$ 排放。1.3 节介绍了环境保护问题及其对核能可持续发展的促进作用。

## 1.1　能源需求趋势

很多国家对未来能源结构进行了前景展望。短期来看，预计全球能源需求不会增长。但中长期来看，世界人口增长、人均收入和生活质量提高等因素将促使绝大多数国家的能源需求增加。因此，全球能源需求将不可避免地增长，并且主要集中于电力需求[1]。在提供大量电力输送的同时，又能够避免化石燃料燃烧的途径之一就是应用核能。

目前，全球有 16 个核能应用大国，其核电占比超过本国电力供给的 20%（图 1 - 1）。欧洲 29 个国家（欧盟（European Union，EU）27 国＋瑞士＋英国）中有 15 个国家应用了核电，共有 132 个核电机组，占总发电量的 27%，贡献了 50%的低碳电源。法国核电贡献了全国总发电量的 75%。本届法国政府的战略是降低核电份额以促进可再生能源发展。中国计划到 2025 年，核电装机容量达到 70 GWe 左右。从中长期

（2035年、2050年）发展趋势来看，中国的电力需求总量持续增长，电能占终端能源消费比重不断提高，电源结构低碳化转型和清洁能源发展日益加快。煤炭目前是中国主力能源，也是造成空气污染和温室气体排放的主要来源。中国正在实施绿色、低碳的能源战略，未来会降低煤炭消费增长速度，并使煤炭消费总量尽早达到峰值，此后总能源的增量将由清洁能源补充。积极发展核电是中国长期重大战略选择，核电可以成为中国能源的绿色支柱。法国的电源结构主要由核电和水电构成，几乎完全脱碳。

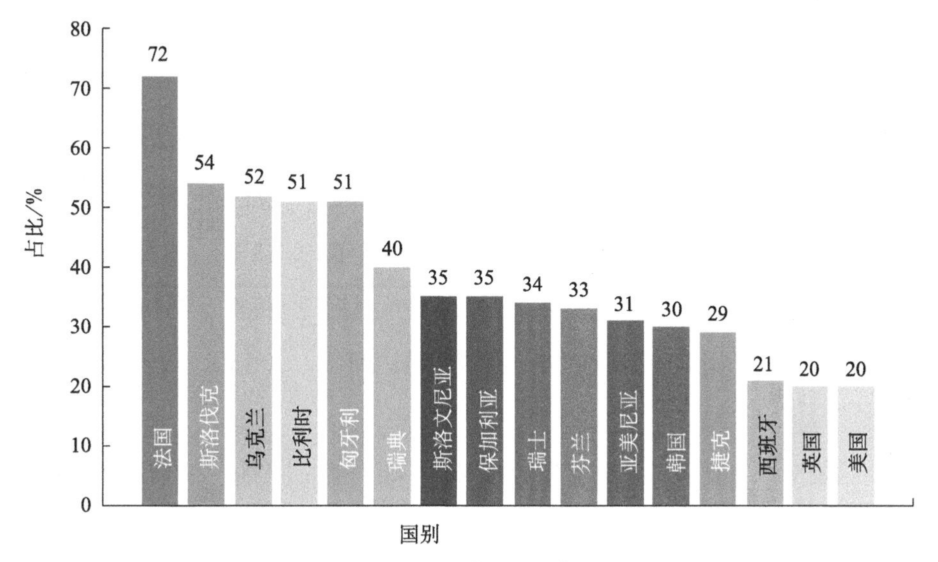


图1-1 核电占比超过本国电力供给20%的国家（2016年12月）

世界能源消费以往由发达国家主导，现在开始向发达国家与发展中国家共享能源市场的格局发展过渡。资源和环境制约、温室气体导致全球气候变化等因素，对传统的世界能源格局提出了挑战，化石能源供给只能保持低速增长。因此，能源生产和利用将进一步向高效、清洁、低碳的方向发展。在今后几十年内，世界能源结构将发生重大变化，进入能源结构多元化（油、气、煤、可再生能源、核能多方鼎立）的格局。

在多元化能源格局中，核能具有消耗资源少、单位发电量的温室气体排放少、土地占用面积小、可调峰和稳定等优点。本报告旨在综合分析核能的环境影响，均衡评估核能的优势和劣势，从而衡量其是否具有长期可持续性，以及能否满足不断增长的电力需求。

## 1.2 脱碳承诺，以及不同发电能源 $CO_2$ 排放情况

1992 年，《联合国气候变化框架公约》（*United Nations Framework Convention on Climate Change*，UNFCCC）呼吁将温室气体浓度保持在稳定水平，防止气候变化风险。各缔约方在《巴黎气候变化协定》中承诺目标是，将全球平均气温较前工业化时期上升幅度控制在 2 ℃以内，并努力将温度上升幅度限制在 1.5 ℃以内。联合国宣布《巴黎气候变化协定》于 2016 年 11 月 4 日正式生效，该协定将为 2020 年后全球气候变化行动作出安排。法国于 2016 年 6 月 15 日正式批准《巴黎气候变化协定》，成为第一个批准该协定的工业化国家。中国于 2016 年 9 月 3 日批准加入《巴黎气候变化协定》，成为第 23 个完成批准协定的缔约方。该协定将对能源格局产生深远影响，促使全球转而采用低碳能源结构。而核能是达到低温室气体排放和实现气候目标的重要途径。

当今核能发电主要采用铀原子裂变技术。1 kg 核燃料（铀-235）裂变释放的能量相当于 2700 t 标准煤燃烧释放的能量，因而属于高效的密集型能源。如果对一个常规核电厂的燃料量和一个燃煤电站的燃料量进行比较，就会更为直观。假定两个电站的装机容量均为 1 GWe 并运行一年，核电厂将使用 30 t 燃料，而燃煤电站则需要 400 万吨煤。在 2.3 节的分析中，本报告指出核能发电不会排放颗粒物且大气污染物也非常少。但在核燃料开采、核电厂建造和核燃料循环的过程中仍难以避免排放 $CO_2$。在全生命周期中，核能发电（包括核电厂和核燃料循环设施）每年排放的 $CO_2$ 仅不到燃煤发电的 1%，也低于太阳能和风能。

## 1.3 环境保护要求核能可持续发展

除了低碳和占地少等优点外，核能还必须具有安全性和经济竞争力。因而，必须评估核电厂和核燃料循环设施的环境影响。此外，还必须对反应堆定期卸出的乏燃料

管理措施、其储存、后处理和放射性废物最终处置的方式等进行详细审查。关于各类设施产生放射性废气、废液和固体废物，气态和液态流出物经处理和储存，达到法规允许水平后可排放到环境中；固体废物在经处理和临时贮存以减少其体积和放射性活度，并符合废物最小化标准的要求后，可进行临时贮存或直接最终处置。

根据安全分析和环境影响评价，核设施允许排放量所致人的剂量水平（放射性和化学毒性）要低于法规允许的规定值。国家法规至少应达到国际原子能机构（International Atomic Energy Agency，IAEA）的要求，但往往会更为严格。目标是达到剂量远低于国际辐射防护委员会（International Commission on Radio Protection，ICRP）推荐的 1 mSv/a 个人剂量限值。法规越来越严格，流出物排放限值越来越低，使得（去污）处理技术持续进步。

在 2017 年的第一期报告中，中法三院对核能在安全性、放射性废物管理、先进核能系统的开发和部署、经济性、公众接受性等方面的问题和挑战进行了分析。相比“核能与安全”议题，“核能与环境”议题没有得到足够充分的讨论。然而，对于工业活动（特别是生产大规模电力的电厂）产生的全球性和地区性环境影响，公众的神经正变得日益敏感。全球性影响更为重要，并且将推动未来能源结构的抉择。在生态转型的过程中，必然伴随着能源转型。

核能是否会造成直接或间接环境影响是当地公众关心的主要社会问题，其将导致他们是否会接受核能。因此，对“核能如何影响环境”进行综合评估就显得格外重要，如果我们对当地公众看待这些问题的方式给予足够的关注，就能够提出全面综合、统筹兼顾的措施，将上述负面影响最小化。希望这一观点能够成为中法三院第一期报告的有益补充，并且对此议题进行更好地评估。核能的可持续性还取决于，核能国家是否相信新兴国家能够遵守环境影响控制原则。

本报告的关键目标是，审议核能对环境的影响并评估其作为清洁能源的潜力。为此，中法三院决定采用多个指标衡量以下情形产生的全球和区域影响：多种情形（正常、事故）下的核电厂及核燃料循环设施的建设、运行和退役过程。同时，对核循环的前端（从采矿到燃料元件制造）和后端（从废物处置到核电厂/设施拆除）每个环节在实施过程中产生的全部影响进行通盘考虑。由于核能生产过程始终存在放射性，因此要首先注意生物过度暴露于电离辐射的风险。

## 1.4 本报告内容和结构

本报告包括综述与建议和 6 章内容。接下来的第 2 章将讨论核电厂和核燃料循环设施正常运行期间的环境影响，包括对各类电力生产系统的温室气体和大气污染物排放进行比较，并讨论了正常运行期间的放射性、用水、土地利用和材料消耗等问题。

第 3 章讨论了乏燃料和放射性废物管理，并介绍了预防环境影响的原则、策略和框架。乏燃料和放射性废物管理一般包括放射性废物分类、处理、排放和处置，以及开放和闭合核燃料循环带来的不同影响。此外，还讨论了各个放射性废物管理环节的环境保护措施。

第 4 章回顾了几次严重的核事故（三里岛、切尔诺贝利和福岛），并强调了从这些事故中吸取的教训。此外，还介绍了现有核电厂的升级改造和新的第三代核电的设计改进，以便限制核场址的边界和减少事故造成的环境影响。

第 5 章介绍了与环境有关的核安全问题；讨论了核安全的目标（限制核事故的可能性、预防措施以及减轻核事故后果）。此外，还探讨了核电厂选址问题、安全监管当局的作用，以及核电厂运营者和政府的责任。

第 6 章总结了主要研究结论。报告的最后附上参考文献和词汇表。

# 第 2 章　核电厂和核燃料循环设施正常运行期间的环境影响

---

建议

对于核活动产生的辐射照射，必须始终将其与天然照射以及其他发电技术产生的辐射照射进行比较。

世界各地大多数的流行病学研究表明，低剂量（率）照射和极低剂量（率）照射的长期影响是无害的。尽管如此，仍建议在新的核设施运行之前开展本底流行病学研究，此举可为事故后流行病学研究的对比分析、辐射危险分析以及回应公众的关切提供有价值的信息。

快中子反应堆和多次循环利用可能会大幅减少核能的环境足迹，因为此二者能够减少铀矿开采活动，并降低核废物的数量和毒性。虽然当下还不需要快中子反应堆技术，但必须做好在未来几十年进行商业化的准备。

---

## 本章介绍

本章讨论了核能的环境足迹，比较了核能发电系统与其他发电系统的环境影响，并探讨了引入减小环境影响的新技术趋势。首先，本章概述了在人类活动框架内关于核能预期环境影响的几个总体考虑。

## 2.1 如何衡量核能的环境影响

环境影响是指“自然”环境特定部分发生的临时或永久性的改变，包括空气、水、土地、植物、野生动物等，而人类自身可能是最终对象。例如，人类活动释放的气体、液体或固体引起的改变都属于环境影响。

### 2.1.1 核能对环境的主要影响

主要的环境影响涉及气候变化（$CO_2$ 和其他温室气体（greenhouse gas，GHG）排放）、空气和水污染（各类释放物）、取水和用水、人造陆地或遗产流失、自然土地退化、土壤侵蚀、原材料消耗，以及废物的产生、处理和处置等。

气候变化被认为对全球环境具有最严重的影响。此外，气候变化还会导致海洋和陆地生态系统退化以及生物多样性丧失。这种退化是由于酸化和富营养化造成的，与硫和氮的气态氧化物以及 $CO_2$ 的排放密切相关。核能发电对此类排放的贡献非常小；核能发电对土地占用、水循环等其他影响的贡献，取决于核燃料循环的方案，具体内容将在后文进一步探讨。

核能的释放物可能具有放射性，它通常会导致人和其他生物受到辐射照射，必须从人类健康和生物多样性角度来评价这类累积照射的潜在影响。为了进行此类评价，必须始终将核能的辐射照射与天然辐射源的辐射照射或医疗诊断的辐射照射进行比较。

(1) 天然本底辐射照射，例如：法国居民受到的天然本底辐射年均有效剂量为 2.9 mSv[2]，美国和中国居民均为 3.1 mSv[3,4]；某些高本底辐射地区的剂量水平可能要高得多，如印度喀拉拉邦的居民受到的天然本底辐射年均有效剂量超过 10 mSv；纽约—巴黎往返航班乘客受到的辐射照射约 0.05 mSv[5]。

(2) 人为活动引起的其他天然辐射源照射，例如：稀土开采和加工，或由燃煤发电导致的辐射照射。由于大量使用煤灰渣作为建筑材料，导致中国室内氡照射显著增强[4]。

(3) 医疗诊断辐射照射的平均值约为 1 mSv/a（文献 [5] 数据取整）。

### 2.1.2　环境释放

环境释放分为两种类型：

(1) 核电厂和设施的放射性物质直接释放。在风和雨等外力驱动下，导致放射性物质被弥散、稀释、向土壤沉积、从土壤中浸析、在土壤中迁移。

(2) 废物包中放射性物质的长期释放（浸出或整备材料的分解）。在天然岩石圈梯度力（水力、热力、化学）作用下，放射性物质以其原本形态或胶体形态迁移。

### 2.1.3　评价

根据所考虑的时间和尺度的差异，可通过两种方法评价大型能源生产系统的环境影响。

当我们评价大的潜在影响时（比如涉及气候变化和长期持续的影响），建议采用"全生命周期分析"（life cycle analysis，LCA，"从摇篮到坟墓"）方法来评估核设施从开始建造到最终拆除（大约一个世纪）对全球造成的影响。LCA 可以充分说明已经记录的和预期的影响。例如，LCA 的评估结果能够对 2.2 节和 2.4 节的数据提供支持。

当影响仅限于设施厂址及其周围，且影响时间仅涉及"日常生活"（即局部影响）时，即时影响和远期延迟的影响均非常重要；但只有前者是可以测量的，后者则需通过模拟来获得。下面的 2.3 节将举例说明这一方法。

在某些情况下，"天然类似物"可为长期影响提供实验数据，例如：放射性核素数十亿年来在奥克洛（Oklo）铀矿床（加蓬）中的有限迁移[6]，或数千年来地中海玻璃表面蚀变的减少——这种现象保护这些玻璃免于溶解。

在评价过程中，需考虑释放物中存在的全部天然和人工放射性核素与其他元素。

核仪器能够探测和鉴别极低水平的放射性，但在化学毒性方面则少有这种情况。无论个人还是组织均可轻易获得基本的低成本核仪器，并经短暂、适当的培训后自行进行就地或远程测量，且能够保证适当的质量水平。因此，放射性测量正日益成为一个可以进行独立研究的领域。在确保独立性的同时，人们可对环境中的放射性水平和性质进行交叉检查。

环境微量化学毒素测量工具要比放射性测量工具复杂得多，因此很难获得化学毒性的就地测量数据。

法国和其他拥有核设施的欧洲/非欧洲国家定期对国内核能、工业或医疗设施中现有的材料、源和废物进行了详细分析，并制定了短期和长期管理规划。据此，可对实际的或潜在的放射性释放作出评价。

### 2.1.4 方法

在预期产生环境影响的每个相关领域中，可选择某些参数来衡量各类潜在的或实际的有害影响，以比较不同能源系统的环境影响。

可根据能源生产情景和能源系统各自的特性来实施 LCA。

在考虑了相关设施和对象的特征以及当地人群的生活方式之后，可以估算局部影响。

可通过模拟方法估算人类或生物群落受到的辐射影响和化学影响。所有程序都应遵循大致相同的步骤并依赖大型数据库提供佐证。然而，此类方式还需要经过充分验证，一些结果可能会存在争议。长期（数千年）影响的模拟结果可能更具争议。

## 2.2 核能的流出物、辐射影响及解决方案

### 2.2.1 核电厂的流出物及辐射影响

法国核电厂正常运行的辐射影响见表 2－1[7]。估算的剂量水平远低于前文提到的天然辐射照射水平（约低 3 个数量级）。

表 2－1 基于设施实际排放量计算的 2011 年以来法国核电厂所致关键居民组辐射剂量

| 法国电力集团（Electricité de France，EDF）核电厂 | 距离/km | 估算的剂量/（mSv/a） | | | | | |
|---|---|---|---|---|---|---|---|
| | | 2011 年 | 2012 年 | 2013 年 | 2014 年 | 2015 年 | 2016 年 |
| EDF / Belleville-sur-Loire | 1.8 | $8\times10^{-4}$ | $8\times10^{-4}$ | $7\times10^{-4}$ | $4\times10^{-4}$ | $5\times10^{-4}$ | $4\times10^{-4}$ |
| EDF / Blayais | 2.5 | $6\times10^{-4}$ | $2\times10^{-4}$ | $2\times10^{-4}$ | $6\times10^{-4}$ | $5\times10^{-4}$ | $5\times10^{-4}$ |

续表

| 法国电力集团（Electricité de France，EDF）核电厂 | 距离/km | 估算的剂量/（mSv/a） | | | | | |
|---|---|---|---|---|---|---|---|
| | | 2011 年 | 2012 年 | 2013 年 | 2014 年 | 2015 年 | 2016 年 |
| EDF / Bugey | 1.8 | $8\times10^{-4}$ | $5\times10^{-4}$ | $6\times10^{-4}$ | $2\times10^{-4}$ | $2\times10^{-4}$ | $9\times10^{-5}$ |
| EDF / Cattenom | 4.8 | $3\times10^{-4}$ | $3\times10^{-3}$ | $5\times10^{-3}$ | $8\times10^{-3}$ | $7\times10^{-3}$ | $9\times10^{-3}$ |
| EDF / Chinon | 1.6 | $5\times10^{-4}$ | $5\times10^{-4}$ | $3\times10^{-4}$ | $2\times10^{-4}$ | $2\times10^{-4}$ | $2\times10^{-4}$ |
| EDF / Chooz | 1.5 | $1\times10^{-3}$ | $9\times10^{-4}$ | $2\times10^{-3}$ | $7\times10^{-4}$ | $6\times10^{-4}$ | $6\times10^{-4}$ |
| EDF / Civaux | 1.9 | $7\times10^{-4}$ | $9\times10^{-4}$ | $2\times10^{-3}$ | $8\times10^{-4}$ | $9\times10^{-4}$ | $2\times10^{-3}$ |
| EDF / Cruas | 2.4 | $5\times10^{-4}$ | $4\times10^{-4}$ | $4\times10^{-4}$ | $2\times10^{-4}$ | $2\times10^{-4}$ | $2\times10^{-4}$ |
| EDF / Dampierre-en-Burly | 1.6 | $2\times10^{-3}$ | $1\times10^{-3}$ | $9\times10^{-4}$ | $4\times10^{-4}$ | $5\times10^{-4}$ | $5\times10^{-4}$ |
| EDF / Fessenheim | 3.5 | $8\times10^{-5}$ | $1\times10^{-4}$ | $1\times10^{-4}$ | $4\times10^{-5}$ | $4\times10^{-5}$ | $3\times10^{-5}$ |
| EDF / Flamanville | 0.8 | $2\times10^{-3}$ | $6\times10^{-4}$ | $7\times10^{-4}$ | $5\times10^{-4}$ | $2\times10^{-4}$ | $2\times10^{-4}$ |
| EDF / Golfech | 1 | $8\times10^{-4}$ | $7\times10^{-4}$ | $6\times10^{-4}$ | $2\times10^{-4}$ | $3\times10^{-4}$ | $3\times10^{-4}$ |
| EDF / Gravelines | 1.8 | $2\times10^{-3}$ | $4\times10^{-4}$ | $6\times10^{-4}$ | $8\times10^{-4}$ | $4\times10^{-4}$ | $4\times10^{-4}$ |
| EDF / Nogent-sur-Seine | 2.3 | $8\times10^{-4}$ | $6\times10^{-4}$ | $1\times10^{-4}$ | $5\times10^{-4}$ | $4\times10^{-4}$ | $7\times10^{-4}$ |
| EDF / Paluel | 1.4 | $8\times10^{-4}$ | $5\times10^{-4}$ | $9\times10^{-4}$ | $9\times10^{-4}$ | $4\times10^{-4}$ | $3\times10^{-4}$ |
| EDF / Penly | 2.8 | $1\times10^{-3}$ | $6\times10^{-4}$ | $7\times10^{-4}$ | $4\times10^{-4}$ | $4\times10^{-4}$ | $4\times10^{-4}$ |
| EDF / Saint-Alban | 2.3 | $4\times10^{-4}$ | $4\times10^{-4}$ | $4\times10^{-4}$ | $2\times10^{-4}$ | $2\times10^{-4}$ | $3\times10^{-4}$ |
| EDF / Saint-Laurent-des-Eaux | 2.3 | $3\times10^{-4}$ | $2\times10^{-4}$ | $2\times10^{-4}$ | $2\times10^{-4}$ | $1\times10^{-4}$ | $1\times10^{-4}$ |
| EDF / Tricastin | 1.3 | $7\times10^{-4}$ | $7\times10^{-4}$ | $5\times10^{-4}$ | $2\times10^{-4}$ | $2\times10^{-4}$ | $2\times10^{-4}$ |

对中国 6 座压水堆核电厂（pressurized water reactor，PWR）和 1 座重水堆核电厂（heavy water reactor，HWR）运行期间的气态和液态流出物的监测结果进行了分析。图 2－1 呈现了 2011～2013 年这 7 座核电厂各类放射性流出物的平均排放量，其

中最大排放量得到了有效的管控。在任何情况下，均远低于监管限值和天然辐射照射水平。文献［8］中估算了2011～2013年中国核电厂流出物排放所致公众的归一化集体剂量，约为$6.4\times10^{-2}$人·Sv/GWa（人·希沃特每吉瓦年）。

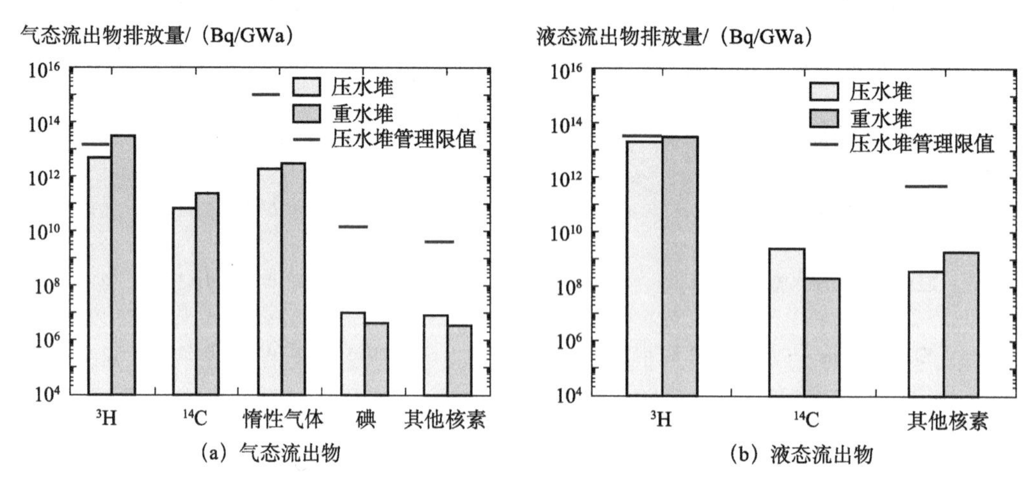


图2-1　中国核电厂平均放射性流出物排放量（2011～2013年）[8]

需要指出，氚（T）是核电厂向环境释放的放射性核素之一。作为氢的一种同位素，氚在环境中的行为主要与水循环（HTO或氚化水）有关，也与光合作用（与植物中$T_2$或HT分子结合）以及有机氚分子在生物体内的代谢（有机结合T或OBT）有关。世界卫生组织（World Health Organization，WHO）建议，氚在长期使用的饮用水中的指导水平为10000 Bq/L[9]。法国辐射防护与核安全研究院（Institut de Radioprotection et Sûreté Nucléaire，IRSN）的报告[10]指出，在法国进行了数十年的研究后，没有证据表明氚能够在植物成分中形成生物积累；对陆地动物产品，结论也是如此，但这一结论基于有限的数据[11]。这是因为多数的研究主要集中于描述氚在动物体内行为的生理模型，用以估算动物产品（奶、肉等）中的氚浓度；然而，转移因子的新近估算值[11]总是小于1，这证实了源自动物的食物中并未出现氚积累。

同样，也没有发现氚在海洋物种中发生生物积累的证据[12]。

多年来，如图2-2所示，在中国和法国，液态和气态流出物排放量都已大幅度减少。

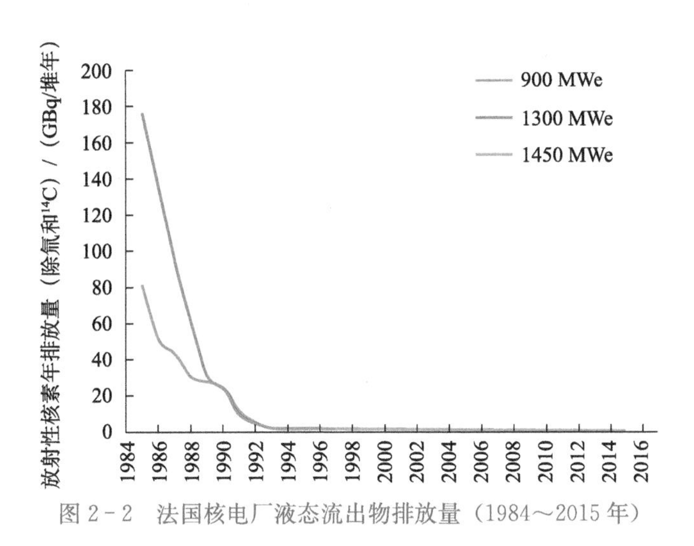

图 2－2　法国核电厂液态流出物排放量（1984～2015 年）

### 2.2.2　核燃料循环的流出物及辐射影响

核燃料循环包括核燃料生产、加工、贮存和后处理等活动。法国核燃料循环设施正常运行的辐射影响见表 2－2[7]。据估算，辐射剂量仍然很低，比天然辐射照射的年剂量低 2～4 个数量级。

表 2－2　基于实际排放量计算的 2011 年以来法国核燃料循环设施所致关键居民组辐射影响

| 核燃料循环设施 | 距离/km | 估算的剂量/（mSv/a） | | | | |
|---|---|---|---|---|---|---|
| | | 2011 年 | 2012 年 | 2013 年 | 2014 年 | 2015 年 |
| Andra / CSA | 2.1 | $3\times10^{-4}$ | $1\times10^{-5}$ | $1\times10^{-6}$ | $2\times10^{-6}$ | $2\times10^{-6}$ |
| Andra's Manche 处置场 | 2.5 | $6\times10^{-4}$ | $4\times10^{-4}$ | $4\times10^{-4}$ | $3\times10^{-4}$ | $2\times10^{-4}$ |
| Areva NP in Romans F | 0.2 | $6\times10^{-4}$ | $6\times10^{-4}$ | $5\times10^{-4}$ | $3\times10^{-4}$ | $3\times10^{-4}$ |
| Areva / La Hague | 2.8 | $9\times10^{-3}$ | $9\times10^{-3}$ | $2\times10^{-2}$ | $2\times10^{-2}$ | $2\times10^{-2}$ |
| Areva / Tricastin | 1.2 | NA | $3\times10^{-4}$ | $3\times10^{-4}$ | $3\times10^{-4}$ | $3\times10^{-4}$ |

中国核燃料循环的流出物排放得到了良好的控制且记录详实。2011～2013 年，中国核燃料循环的流出物排放量及所致公众辐射剂量分别见图 2－3 和图 2－4。

此外，中国对 2011～2013 年核能发电全生命周期（LCA）所致公众归一化集体

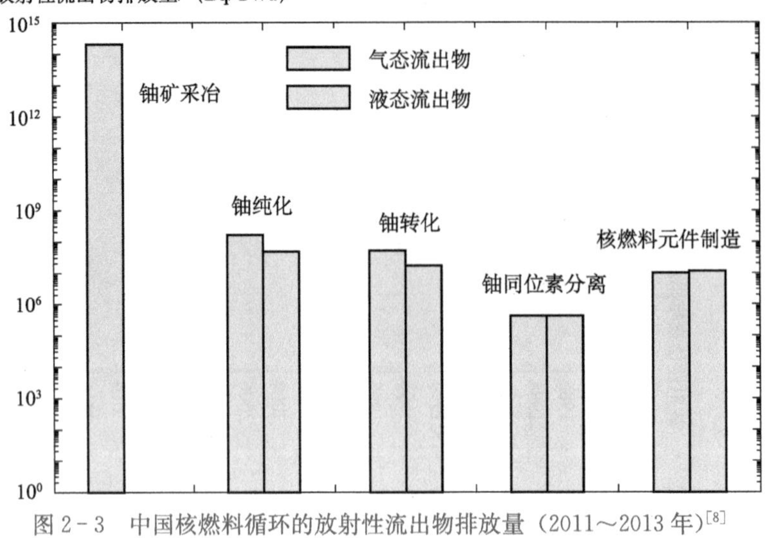


图 2-3　中国核燃料循环的放射性流出物排放量（2011～2013 年）[8]

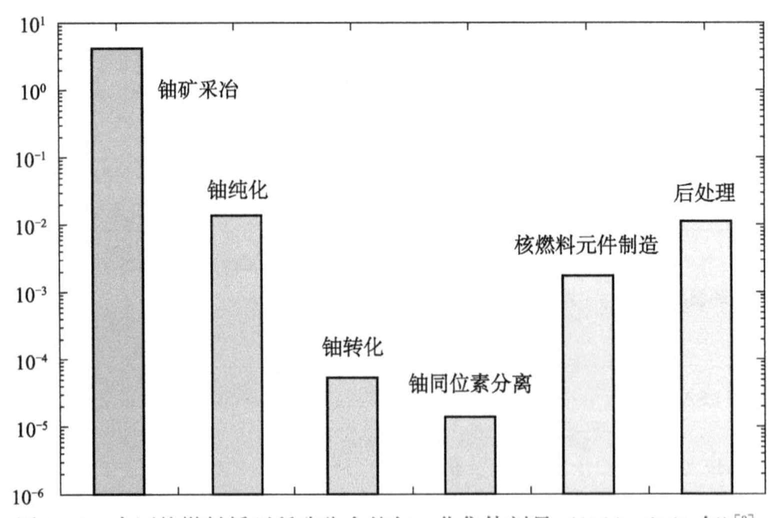


图 2-4　中国核燃料循环所致公众的归一化集体剂量（2011～2013 年）[8]

（铀矿开采和水冶数据基于$^{222}$Rn 的排放，其他数据基于总铀的排放）

剂量的估计值约为 4.6 人·Sv/GWa[7]（评价范围为厂址周围半径 80 km，详见附录 2-1，同时包括其他发电技术的辐射环境影响），其中 86%来自于铀矿开采和水冶。在不久的将来，随着就地浸出铀矿冶技术的持续推广，中国核能发电链所致公众的剂量水平将会进一步降低。

### 2.2.3　辐射监测和监控

在所有的核能国家，环境放射性监测是所有营运单位、安全和环境当局的主要关切。本节主要介绍法国和中国的监测系统。

法国通过三个远程监控网络实施放射性监测：

(1) 核设施的所有者/营运单位在厂址附近（法国规定 0～10 km）建设和运行的取水（水监测）和信标（空气监测）系统。

(2) 空气放射性监测网络，其目的是及时发现空气中放射性水平的任何异常升高。在法国，环境 γ 剂量率监测网（Teleray）包括分布在全国范围内的 400 个信标，并特别关注距离核设施厂址不到 30 km 的大城市。

(3) 在法国七条主要河流的河口上游或其流入邻国入口处布置水中放射性连续监测网络：探测限极低，$^{137}Cs$、$^{131}I$ 和 $^{60}Co$ 的探测限在 0.5～1 Bq/L 的范围内。

此外，还建立了采样监测网络以评估所有利用放射性核素的人类活动对环境空气的影响：由法国 IRSN 运行的采样监测网络（OPERA-AIR）配有 40 个 OPERA-AIR 台站，其中 32 个设置在核设施厂址附近，用于采集水、底泥和沉积物以及奶样品。

中国的监测系统由多层次的环境辐射监测网络构成，用以监测核设施运行期间的环境辐射水平。

(1) 近距离（一般在距核电厂址半径 5 km 范围内）：固定式自动监测站点，由营运单位建设、运行和管理（自主监测）。

(2) 远距离（一般在距厂址半径 20 km 范围内，不包括厂区）：固定式自动监测站点，由营运单位建设，并由所在地省级环境保护主管部门运营管理（监督性监测）。通过自主监测和监督性监测的结合，能够对环境 γ 辐射水平以及大气环境介质样品分别进行测量。

(3) 厂区外围（一般为距厂址半径 10 km 范围内，包括厂区），营运单位对地表水、地下水、受纳水体、土壤、底泥和生物等环境介质样品进行监测和采样分析。

对于全国范围内主要城市和区域，由国家辐射环境监测网点对大气、水体和土壤等环境介质进行监测和采样分析。

### 2.2.4 电离辐射的生物效应

超过阈值的高剂量辐射照射会产生确定效应；低剂量（率）的照射情况可能引起随机效应，一般为癌症或遗传疾病。

为探讨低剂量（率）电离辐射照射对大规模人群可能造成的健康影响，中国从1972年开始在广东阳江市天然放射性高本底地区（high background radiation area，HBRA）与正常本底辐射对照地区（control area，CA）开展了调查研究工作，其中HBRA和CA居民受到天然辐射照射的年均剂量分别为6.4 mSv和2.4 mSv（在2000年，将其分别修订为5.9 mSv和2.0 mSv）。

根据对HBRA累计1008769人年和CA累计995070人年癌症死亡率的调查，以及对HBRA的13425人和CA的13087人的遗传性疾病、先天性畸形检查、外周血淋巴细胞的染色体分析和免疫功能的测定结果，未观察到天然辐射对HBRA居民健康产生有害影响。

自20世纪50年代以来，欧美国家也在核设施厂址周围进行了流行病学调查。结果表明，与对照区相比，核设施附近居民的癌症死亡率和儿童白血病发病率未发现显著差异。这主要是因为，核设施正常运行期间放射性物质排放所致公众辐射剂量非常低，关键居民组受到的附加剂量约为10 μSv/a，约占除氡外的天然本底辐射水平（约1 mSv/a水平）的1%。

在低剂量电离辐射条件下，癌症诱发和生物效应的机理分析，以及核设施周围公众的剂量评价均具有较大的不确定性，因而流行病学调查难以给出辐射危险的定量结论，也就没有必要开展大规模流行病学常规调查。虽然基于辐射影响生化特征的新方法值得期待，但这类方法尚未投入使用。

但是，在新的核设施运行前开展流行病学本底调查能够为事故后的比较分析提供有价值的信息。这不仅能够帮助评价辐射危险，并有助于对核设施公众关注的因事故期间及事故后放射性物质的最终释放而导致的较高照射剂量的后果作出回应。

附录2-2中提供了更多关于流行病学调查的信息。

### 2.2.5 放射性物质的运输

为了满足工业、医疗或研究机构的需求，法国每年要运输约90万个放射性货包，其中大部分为极低水平的放射源和废物货包。仅 15%与放射性燃料以及低、中或高放废物有关。全世界范围内每年运输的放射性货包数量达 1000 万件，这仅占所有有害物质运输货包总数的 2%。

对于人类和环境来说，放射性物质运输的主要风险是辐射和污染。法国每年会发生 1～2 起导致放射性物质向环境释放的运输事故，但影响均十分有限，最严重的情况可探测到微弱的污染，但已通过局部去污操作进行了处理。

对于重件或大件货包而言，铁路运输具有非常高的安全水平，通常予以优先考虑。

海路运输约占放射性物质运输总量的 4%，主要用于新燃料、乏燃料以及高放废物的运输。运输船舶会根据国际海事组织的要求专门设计。

公路运输是最灵活的放射性物质运输方式。公路运输需要遵守特定的规则，以避开拥挤时段和居民区。

航空运输仅用于小件和紧急货包（例如放射性药物），以及长距离运输。

应根据放射性物质的特点和运输要求选用适当的运输方式。

### 2.2.6　利益相关方参与

法国为那些被归类为对环境保护重要的、最危险的设施（Installation Classée Pour l'Environnement，ICPE）设立了地方利益相关方信息委员会（Commission Local d'Information，CLI）。

法国共设立了 53 个 CLI，其中 38 个位于核设施厂址附近。CLI 共拥有约 3000 名成员，包括当地政治家、工会代表、协会代表、专家和有资质的人员。其总体任务是向公众通报 ICPE 设施的安全及其对人员和环境的影响。在核领域，《透明与安全法》(2006 年 6 月 13 日）为他们提供了法律依据（《环境法典》第 125 - 7 条)。

CLI 全国协会（Association Nationale des CLIs，ANCCLI）负责汇总 37 个 CLI 的经验和诉求，并将他们的集体见解提交国家和国际机构予以关注。

在利益相关方参与方面，中国建立了“中央督导、政府主导、企业作为、社会参与”的公众沟通工作机制，以促进“科普宣传、公众参与、信息公开、舆情响应、融合发展”工作。

《中华人民共和国核安全法》为公众在重大核能项目中的知情权、参与权和监督权提供了法律依据和保障；涉核重大新项目建设纳入地方人大审议制度，涉核重大项目公众沟通纳入地方社会管理体系；各涉核单位需制定公众沟通策略和中长期规划，把公众沟通工作纳入单位的运作管理；此外，由专业、权威的第三方进行同行评议（包括社会组织、大学和智库，例如中国核能行业协会、中国核学会、中国科学技术协会、中国环境保护协会等）。

## 2.3　核能发电与其他能源发电的环境影响比较

电力生产的环境影响取决于发电技术种类。

在向脱碳经济过渡的框架内，本节的目标是研究各类发电技术造成的非放射性环境影响，例如温室气体（GHG）排放、土地占用、建筑材料消耗、水资源消耗、退役废物管理（核能的辐射影响详见 2.2 节）。这些数据大多来自全生命周期分析（LCA）。附录 2－1 总结了中国各类发电技术全生命周期温室气体排放和辐射影响的最新分析结果。

图 2－5[13] 表明，化石燃料发电厂每生产单位 MWh 电能的 $CO_2$ 排放量，要比核能、风能、太阳能和水力发电的排放量约高 1～2 个数量级。

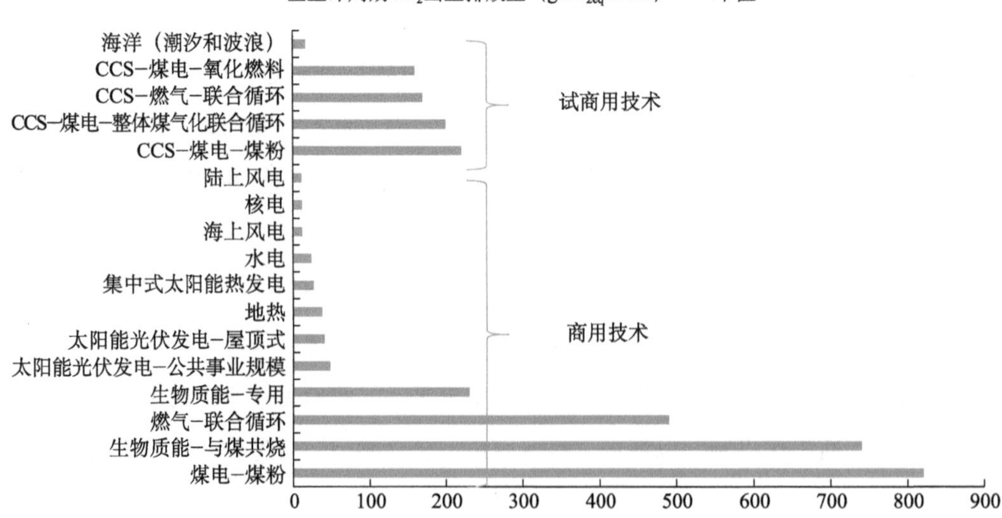


图 2－5　不同发电技术全生命周期 $CO_2$ 当量排放量

按中值（$gCO_{2eq}$/kWh）递增排列，图中 CCS（carbon capture and storage）指“碳捕获和封存”

核能对 $SO_x$ 和 $NO_x$ 气体排放的贡献（约 20 kg/MWh）分别比化石燃料和光伏发电低 2 个数量级和 1 个数量级，因此对土壤酸化和水体富营养化影响较小。水力和风力发电产生的 $SO_x$ 和 $NO_x$ 排放均低于 10 kg/MWh。

风能、太阳能、核能以及水力发电都有其自身的环境影响。必须考虑下列影响：

（1）土地占用；

（2）建筑材料消耗；

（3）水资源消耗；

（4）退役废物。

当作为间歇性电力来源的风能和太阳能设施处于不可用状态时，必须使用化石燃料设施作为后备，因此也应为其分配化石燃料设施的 $CO_2$ 排放份额。

### 2.3.1　土地占用

许多研究和调查项目已经阐述了能源系统造成的土地影响，并日益关注“可再生能源”。表 2－3[14, 15] 根据这一知识简要总结了不同能源系统单位装机容量（MW）的土地占用情况（LCA：全生命周期分析）。

表 2－3　不同能源系统单位装机容量的土地利用强度

| 能源技术 | $m^2$/MW | 能源系统边界<br>能源提取区域及电厂厂址 |
|---|---|---|
| 水电：水库 | 20000～10000000 | 发电机和水库厂址 |
| 太阳能光伏发电 | 10000～60000 | 光伏系统厂址，包括太阳能聚集区。现有建筑物上的光伏系统基本上不会造成土地占用面积的净增加 |
| 太阳能热力发电 | 12000～50000 | 集中式太阳能热电系统厂址，包括太阳能聚集区 |
| 风电 | 2600～1000000 | 低值仅适用于厂址，包括风机及通道的占用面积。高值包括风机之间的土地，通常可用作耕地或牧场 |
| 核电 | 6700～13800 | 低值仅限于厂址。高值包括输电、供水和铁路线路，但不包括采矿、加工或废物处置用地 |

水电的土地占用面积数据看似较高，然而，在很多情况下发电只是大坝的多种用途（灌溉、生活和工业用水蓄水、航运、防洪）之一。水库用地仍未超出合理需求范围。在能源供应方面，建立水库的目的不仅是输送电力，而且还能灵活储存电力，从而创造附加价值。

核电厂的直接用地非常少。因此，核能在土地利用方面具有优势，能够保护生物多样性不会因土地占用和人类活动而受到破坏。

需要指出，全球对核能的认知仍然受到诸如切尔诺贝利和福岛这类事故产生的足迹的影响（见第 4 章）。因此，在发生严重核事故后，公众舆论要求对放射性区域的用地设限就显得合情合理。导致的结果之一就是要能够将事故影响局限在核场址范围内，从而避免任何撤离的行动（见第 4 章）。核事故具有局部的负面外部影响，在采取后续行动时要格外谨慎。需要指出，温室气体排放关乎整个地球的命运，是一种非局部的负面外部影响。

### 2.3.2 建造材料

在相同的装机容量下，核电厂的土建工程要比燃煤发电厂或联合循环燃气轮机发电厂（combined cycle gas turbine，CCGT）需要更多的混凝土和钢材：前者约为 600 t/MW，而后者约为 10 t/MW。这是由于核电厂厂房的安全等级极高，设计必须足以抵御强震，且拥有安全壳系统、防止飞机坠毁的保护壳和复杂的混凝土筏基等；此外，在筏基中还设计了一个可对意外熔化的堆芯进行捕集和冷却的抑制槽（堆芯捕集器）。其中一些特性是第三代核电厂特有的。如果我们考虑一个核电厂在 60 年使用寿期内所生产的大量电能带来的效益，混凝土和钢的消耗量也就不那么重要了。

陆上风电场的负荷因子相对较低，因此生产单位 MWh 电能要比核电厂需要更多的混凝土。如果选择重力式基础，海上风电场就需要更多的水泥、骨料和堆石料用于海底地基的建造。目前，关于漂浮式海上风电场的经验太少，无法提供合理的锚碇基础需求量数据，但其骨料和堆石料的需求量将远低于在海基上架设风桅杆的所用量。

太阳能光伏发电厂以及集中式太阳能发电厂也需要钢筋和混凝土。这两种电厂都需要钢筋混凝土板和钢支架。由于负荷因子较低，电力集中度不高，导致每生产单位 MWh 电能的材料需求量相对较高。

场址内的连接系统以及场址与电网的连接系统对铜和铝有很大的需求。特别是那些必须与海岸相连的海上风力发电场，对铜的需求很高。

基于 2.3.1 节中提到的原因，对于水电站骨料和水泥用量的估算结果并不十分有价值，因为这类需求的多寡，取决于坝址附近是否有足够多的岩石和骨料。许多大型水坝都采用土石坝设计，因为这是最具性价比的解决方案。然而，土石坝对混凝土需求量仍然很大，而且会随着最大洪水的规模和装机功率而进一步增加。

根据多份报告[15, 16] 的研究成果，这一因素造成的每生产单位 TWh 电量的材料消耗量（基于全生命周期分析）见图 2－6。

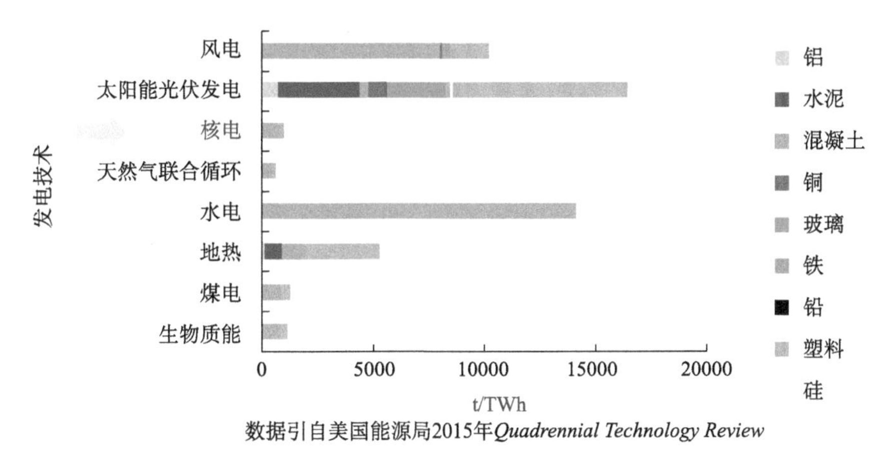


图 2－6　不同发电技术所需的材料量（不包括燃料）

## 2.3.3　取水和用水

核电厂需要大量的水来冷凝驱动主汽轮机的蒸汽。卡诺定律适用于所有的热力电厂，反应堆产生的热能中约 1/3 被转换成电能，余下的 2/3 被排放到环境中。这一问题的环境后果取决于所选择的技术，本部分将进行简要概述。

（1）许多核电厂会建在海边，并利用海水冷却。冷却海水的温度会在冷凝器中升高约 7 ℃，之后重新返回海洋：经初步稀释，表层海水温度升高不超过 1 ℃，这个区域为 1～20 $km^2$ 范围[17]。冷却水流量必须足够大，以确保海水温升不会影响附近水生生物。

（2）若核电厂建在大型内陆河流附近，可使用两种技术方案（见图 2－7）。

● 直流冷却系统：从河流中抽取的冷却水流经冷凝器后返回河中。为确保有限温升，水流量必须足够大。但实际的耗水量不会超过随着温度升高而返回到河里的冷却水的额外蒸发量。虽然需要大量的冷却水，但实际消耗量却少得多。

● 湿式冷却塔系统：一部分用于冷却冷凝器的水被蒸发到大气中，并从其他用水需求中得到补充。

水的潜热远大于显热（蒸发 1 L 水所需能量是将水从 0 ℃加热至 100 ℃所需能量

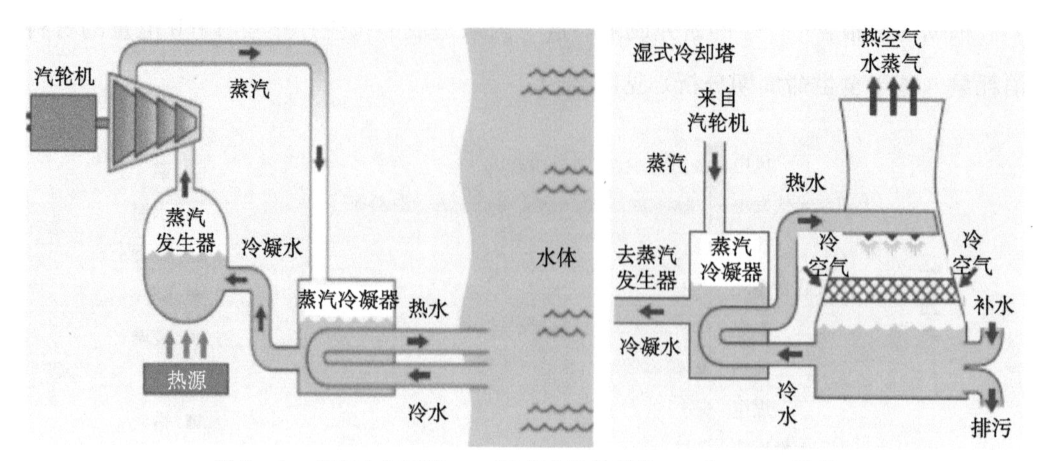

图 2-7　直流冷却系统——湿式冷却塔系统 - 由 SFEN 提供

的 5 倍)，湿式冷却塔比直流冷却的取水量少，但水的损失量较大。

在美国进行的“发电生命周期用水：文献综述和评论”[18] 研究中，发布的以下数据表明，由于选用了不同的发电技术及用水方案，取水量和耗水量有很大差异。

图 2-8 汇总了不同发电技术的耗水量范围①。

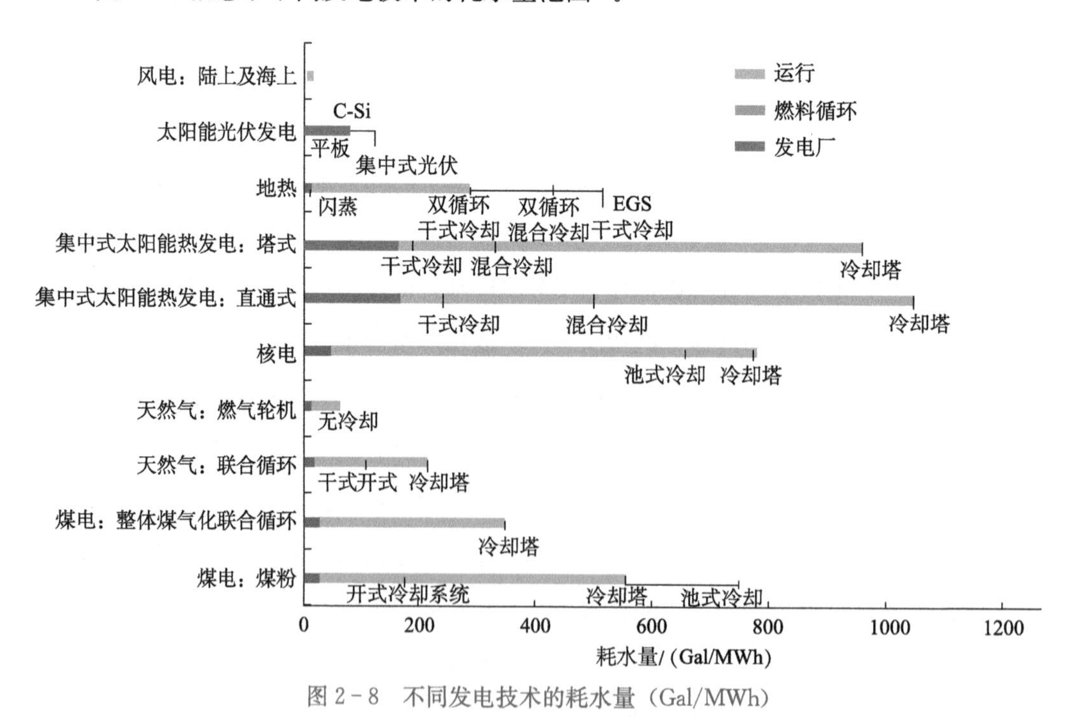

图 2-8　不同发电技术的耗水量 (Gal/MWh)

① 按照通常的用水评估，本文将水的利用分为取水，即“从地下取水或从地表水源中引水”，以及用水，即未返回到周围水环境的那部分取水。

由图2-8可见，风和地热能发电厂的耗水量远小于核电厂。许多集中式太阳能发电厂的耗水量也很小，不过这取决于它们的冷却方案。该图同样显示，核电厂比燃气和燃煤电厂消耗的水更多，因为此二者的热力学效率较低。

图2-9中显示了不同发电技术的取水量范围（开式循环生产每单位MWh电能的取水量超出了该图的范围，未能完全显示）。

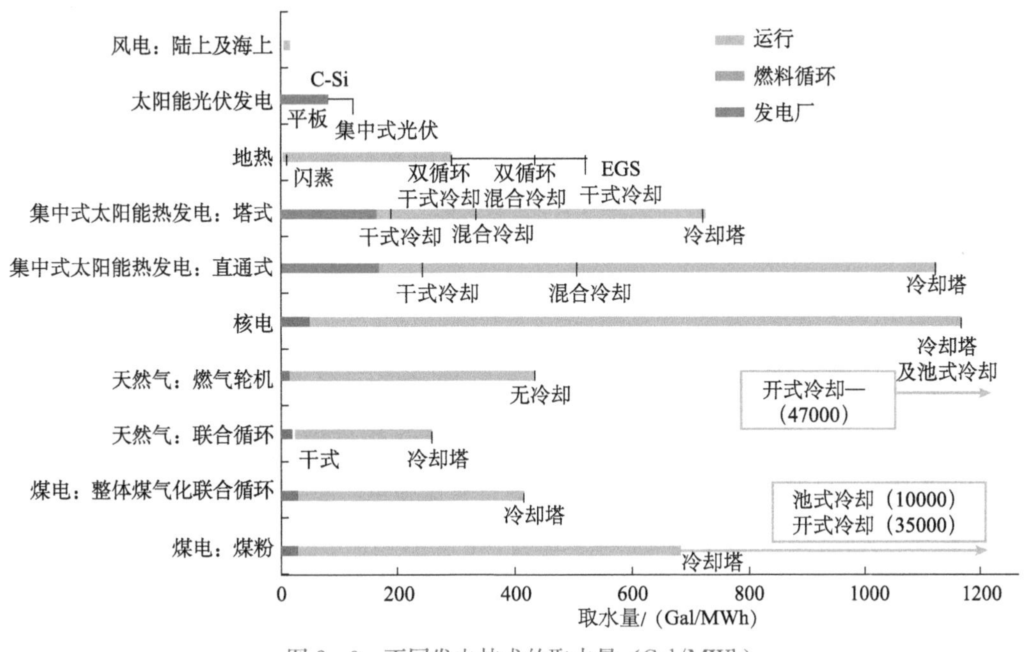


图2-9　不同发电技术的取水量（Gal/MWh）

通常，在法国，发电行业的取水量约占所有行业年总取水量的61%，如表2-4所示；但其中大部分的水仅被具有一次通过冷却系统的6个机组使用，而后返回到河里；采用以上图示数据计算的耗水量适用于所有的压水堆机组，约占所有行业年总取水量的6%。

**表2-4　来源：INSEE/BNPE水统计数据**

| 法国各行业取水量/$Mm^3$ | 2013年 | |
|---|---|---|
| 饮用水 | 5283 | 19% |
| 工业和其他商业用途 | 2745 | 10% |
| 农业 | 2776 | 10% |
| 热电 | 17023 | 61% |
| 合计 | 27827 | 100% |

确实有大量能量被排放到环境中。在法国，只有最早的6个机组采用河水直流冷却系统，当局对其后的所有机组强制采用湿式冷却塔系统。由于冷却水温度较高，导致效率损失，由此造成发电损失约4%。此外，当局还采取了排水温度限制措施，如果电厂向环境排放的水温超过了约束，就会被强制降低功率。然而，其影响仍然有限：从2000年到2017年，因热约束导致的平均产能损失为0.18%；只有在2003年异常炎热的夏季，才达到1.2%的历史最大损失[19]。

法国罗纳河沿岸坐落着14座核电厂，对水温升高的贡献为1.2℃（平均值）或1.6℃（一年中最热的18天），温升仍在合理范围内[20]。但是，内陆电厂的选址必须非常谨慎。电厂可能因冷却用水与其他用户发生用水矛盾，特别是在水资源紧张的地区。

### 2.3.4 常规退役废物

本部分暂不讨论本报告第3章中涉及的放射性废物。但值得一提的是，核能发电产生的非放射性技术废物分别比煤和石油发电少4个数量级和3个数量级。图2-6表明，除铅外，太阳能和风力发电厂消耗的材料比核电厂多10倍到数百倍。按照相对功率因子放大后，可再生能源每生产单位kWh电能比核电厂多20倍到数百倍的材料（如混凝土、铜和铝等）。

还应指出的是，太阳能和风能发电技术的使用寿命均相对较短。这类设施的拆除和重建相对频繁，部分材料的回用问题仍有待解决。

### 2.3.5 关键材料

另一个重要环境参数是材料的稀缺性，例如某些用于太阳能光伏和风能技术的材料（如稀土元素）就比较稀缺，而水电和核电技术几乎不会使用这些材料[15]。作为战略元素的镍是一个例外：核电厂使用了大量的不锈钢，因此也使用了大量的镍。

## 2.4 新技术展望

《巴黎协定》中包含以下目标承诺："在本世纪下半叶实现温室气体源的人为排放

与汇的清除之间的平衡”（第 4 条，2015 年），这个目标通常被称为“碳中和”。大多数的人为 $CO_2$ 排放主要是由煤、天然气和石油等化石燃料的燃烧导致的（约占 87%）[21]。2014 年，电力和热能的生产和输送占燃料燃烧 $CO_2$ 排放量的 70%[22]。因此，必须大幅限制这些行业，甚至停止其使用化石能源。本节论述了减少燃料燃烧的碳排放的主要观点。由于实现这一目标的主要手段之一就是利用核能，因此也介绍了减少核电厂废物的前景。

### 2.4.1　减少燃烧燃料的碳排放

减少燃料燃烧的碳排放的两个主要途径包括：碳捕获和封存（CCS），或直接生产无碳电力。

（1）CCS 技术面临着 3 个挑战：降低成本、提高公众的接受性以及发展封存能力。目前全球只有 40 Mt/a 的 $CO_2$ 的封存能力，根据国际能源署（International Energy Agency，IEA）设定的限制全球长期平均温度上升不超过 2 ℃的情景，至 2040 年应实现 4000 Mt/a 的碳捕获与封存目标（经合组织国家 30%，非经合组织国家 70%）[23]。同样需要强调的是，碳捕获效率最高为 90%，因此，如果不发展碳汇，就不能达到碳中和的目标。总而言之，碳封存只能减少 $CO_2$ 排放，但不足以实现碳中和。

（2）无碳发电既是低碳移动出行的关键，也是低碳能源系统的核心。低碳移动出行可由生物燃料或在电池中储存的电力实现，或通过将氢气转化为电能的燃料电池来产生。

生物燃料是实现碳中和的手段之一，但其局限性亦非常明显：与相同土地利用强度下的太阳能或风能相比，生物燃料产生的电能很少。当世界人口在稳步增长时，生物燃料发电将与粮食生产争夺用地，而粮食生产势必会被优先考虑。因此，风能和太阳能才是生产无碳电力的领先技术。近年来，它们的成本也大幅下降。不过，风能和太阳能也同样受到间歇性限制。

应对间歇性的唯一解决方案是储能。目前，电池可用于日常储能。然而，需要考虑比日常存储更大的容量。电池无法储存在数周甚至数月内平衡电力盈余和缺口所需的大量能源。可以考虑许多替代解决方案：机械（压缩空气、水电储能）、热（熔盐等），但没有一个方案能提供所需的存储能力。

由此，可以得出结论：只有水力发电（但可用厂址不足）和核能发电才有潜力产生可调度的无碳电力。

### 2.4.2 嬗变技术

嬗变是将核燃料循环产生的高放废物最小化的一种方案。主要有 2 种嬗变技术：加速器驱动系统（accelerator driven system，ADS）嬗变和快中子反应堆（fast neutron reactor，FR）嬗变。

包括诺贝尔奖获得者卡洛·鲁比亚（Carlo Rubbia）在内的科学家们提出了 ADS 概念。在该系统中，临界状态是通过添加由散裂产生并加速得到的外部质子源来实现的，从而实现裂变产物的嬗变。但这类技术面临着巨大的技术挑战，且在发电方面是否具备经济竞争力仍然存疑。虽然其具有嬗变锕系元素的潜力，但裂变产物的嬗变将极具挑战性。地质处置库面临的长期风险通常取决于裂变产物，这些裂变产物往往比锕系元素更容易迁移，因此 ADS 的益处——如果其循环能够有效运行——将仍然是有限的[24]。

快中子反应堆基本上可燃烧贫化铀，从而绕过燃料循环的前端（特别是铀矿开采），进一步减少核能系统的环境足迹。此外，还可以控制钚的储备将其扩散的风险降到最低，并控制随后的增殖。第四代核能系统国际论坛（Generation-Ⅳ International Forum，GIF）是研究和开发下一代核能系统的国际合作组织。该论坛鼓励开发 6 种前景良好的反应堆技术，其中有 4 种是快中子反应堆：气冷快中子反应堆（gas cooled fast reactor，GFR）、铅冷快中子反应堆（lead-cooled fast reactor，LFR）、熔盐反应堆（molten salt reactor，MSR，包括快中子熔盐反应堆（fast spectrum molten salt reactor，MSFR））和钠冷快中子反应堆（sodium-cooled fast reactor，SFR）。

与前端活动（矿石采冶、$U_3O_8$ 转化为 $UF_6$、$UF_6$ 富集、$UF_6$ 转化为氧化物等）相比，乏燃料后端循环的影响是有限的。影响最小的是多次循环利用，以及快中子反应堆。

根据法国核设施基地的全生命周期分析（LCA），表 2－5[25] 进行了如下比较：

（1）一次通过燃料循环（once through cycle，OTC）（乏燃料被认为是最终的废物）；

(2) 两次通过燃料循环（twice through cycle，TTC）（乏燃料经一次处理，用于回收铀钚氧化物混合燃料（MOX）中的钚和回收铀燃料（URE）中的铀，目前已在法国部署）；

(3) 第四代快中子反应堆燃料循环（理论上是 100%的钠冷快中子反应堆设计，但易于推广至其他第四代快中子反应堆设计）。

表 2-5　三种燃料循环方案比较

| 影响指标 | 单位 | OTC | TTC | SFR |
|---|---|---|---|---|
| $CO_2$ 排放 | g/kWh | 5.45 | 5.29 | 2.33 |
| $SO_x$ 排放 | g/MWh | 18.73 | 16.28 | 0.59 |
| $NO_x$ 排放 | g/MWh | 29.01 | 25.3 | 3.83 |
| 土地占用 | $m^2$/GWh | 222.6 | 211 | 50.2 |
| 液态化学流出物 | kg/GWh | 333.92 | 287.53 | 12.6 |
| 气态放射性流出物 | MBq/kWh | 0.8 | 1.22 | 0.53 |
| 液态放射性流出物 | kBq/kWh | 2.8 | 27.2 | 3.56 |
| 高放废物（HLW） | $m^3$/TWh | 1.17 | 0.36 | 0.3 |

比较结果清晰地表明，多次循环活动可以改善环境指标。

从上述分析可以看出，第三代（Gen-Ⅲ）和第四代（Gen-Ⅳ）反应堆的线性组合的优势。

实施再循环可大幅度减少高放废物量，从而决定了其高余热能所要求的地质处置库的大小：经再循环后，处置库的体积和面积可缩小到原来的 1/2 以下。

TTC 比 OTC 的放射性气体和液体释放物更多，这是由于在后处理厂中溶解乏燃料造成的，主要来自氪（$^{85}$Kr）和氚。这些放射性释放物远低于监管限值，对健康和环境的影响可以忽略不计。它们的影响低于 10 μSv/a，或低于除氡外天然辐射照射的 1%。

**对开发中的技术进行简要展望**

尽管池式钠冷反应堆提出许多技术问题，但仍然是发展快中子反应堆的首选途径。在不断取得的成就和进展中，值得一提的包括：

(1) BN-800 反应堆是钠冷却的快中子增殖反应堆，建于俄罗斯斯维尔德洛夫斯克州扎列奇内地区的别洛雅尔斯克核电厂，于 2016 年实现商业运营。

(2) 中国计划根据中国实验快堆（China Experimental Fast Reactor，CEFR，热

功率 65 MWth，电功率 20 MWe）的经验开发 600 MWe 钠冷示范快堆。

（3）法国原子能委员会（CEA）完成了 ASTRID（用于工业示范的先进钠技术反应堆）的基本设计，这是一种 600 MW 钠冷反应堆，具有先进的钠气概念，可将能量从反应堆转移到汽轮机。

（4）其他技术也在考虑之中，其中值得一提的是熔盐快堆概念，比如俄罗斯主导的熔盐锕系元素再循环与嬗变堆（MOSART）概念（无论是否有 Th-U 的支持），或法国国家科学研究中心（CNRS）描绘的熔盐快中子反应堆。据称，这些反应堆在必要时能够从铀基（稀缺）逐步转变为钍基（丰富）循环。然而，这一目标备受质疑，因为快中子反应堆本身可以缓解铀短缺的问题，而如果使用钍，将需要投资建设一套全新的燃料循环基础设施。

### 2.4.3 其他先进技术

其他先进技术包括：

（1）新一代耐事故燃料（ATF）。ATF 的设计能够更好地抵御核事故；在正常运行条件下，SiC 燃料包壳有利于燃料冷却，并能够通过铬涂层或 SiC 包壳显著减少锆—水反应产生的氢气，进而限制燃料温度；燃料芯块将被设计为增加导热性并减少放射性释放；并且 ATF 本身对废物的产生没有影响。

（2）人工智能（artificial intelligence，AI）工具。能够防止网络攻击，在将集成到核电厂设备的传感器和进行诊断/监视设备行为的数字耦合算法结合后，将有助于核电厂的运行。

## 2.5 结论

总的来说，核能的环境影响有着良好的记录。

测量和复查环境中的放射性核素的浓度非常容易，而对化学毒性元素浓度的量化则困难得多。

放射性水平实时监测是一种环境保护预警，所获得的数据能够对实际释放或预期释放的模拟结果进行校验。争议之处（如果有的话）在于如何评估相关的危害。对于辐射危害而言，需要将剂量作为附加参数，但这一参数并不适用于化学毒性物质的预防。

核能不会通过燃烧向大气释放化学污染物[5]。对环境的影响来自核燃料循环各阶段（如运输）可能释放的放射性物质。

释放物的放射性水平由辐射防护监管当局规定。根据不同情景下厂址附近最大受照个人或厂址附近的关键人群组所受剂量决定是否授予排放许可。而实际的排放水平可能仅占许可水平的百分之几。还有一些限制要根据最大允许限值（单位为 Bq，根据国际组织确定的水、空气和一些生物的指标计算得出）来确定。导出这些值的情景考虑了所有放射性潜在释放引起的照射，并对应于一年内 1 mSv 的最高待积剂量。

与天然环境放射性、利用煤或天然气等化石燃料的辐射影响相比，核能对现存物种（包括人类在内）局部地区的辐射影响通常较低，甚至极低；事实上，它们低到了无法被识别的程度。核设施排放的化学污染物与其他人为活动排放的污染物的转移情况别无二致。这些评估的依据包括营运单位、管理机构和利益相关方采用多种监测仪器获得的结果，以及流行病学研究结果。

能源系统的全生命周期分析考虑了电厂和设施的建造材料制造，以及运行期间产生的所有环境影响，分析表明：与所有燃烧化石燃料的能源系统相比，核能具有 $CO_2$（和其他温室气体）近零排放的特点，是占用土地和消耗材料最少的能源。在生产相同数量电能的情况下，核能所占用的土地面积要比其他低碳能源低 1～2 个数量级，每 MWh 电力生产所需的混凝土和钢材更少，且不需要稀土元素等关键材料。

由于内陆反应堆冷却的耗水量很大，在水资源紧张的地区进行建造时须谨慎考虑。然而，对于通过海洋或大型河流进行冷却的电厂来说，并不存在严重的冷却水问题。

如果仔细审视建造第三代核反应堆的土建工程，似乎混凝土和钢材的消耗量要高于化石燃料发电厂。这是由内外部事件（地震、飞机坠毁等）的安全要求决定的。

快中子反应堆和多次循环有可能大大减少核能的环境足迹，应加大资金投入，促进这类技术的大力发展。

# 附录 2-1 关于中国的具体情况

在所有的核能国家中，中国的核能发展速度最快。近年来，中国对核电、化石能源（煤电），以及可再生能源（水电、风电和太阳能光伏发电）的温室气体排放和辐射环境影响开展了深入研究。研究采用了全生命周期分析方法（LCA），既考虑了建设和运行期间温室气体和放射性流出物的直接排放，也考虑了能源链系统及相关基础设施所消耗的能源和原材料在其开采、制造、加工和运输期间产生的间接温室气体和放射性流出物排放。设施建造所采用的主要材料在其生产过程中的排放被折算到各发电能源的总排放量中。

不同发电能源全生命周期温室气体的归一化排放量见图 A2-1，核电、水电和风电的温室气体排放水平较煤电低 2 个数量级，太阳能光伏发电的排放水平较煤电低 1 个数量级，处于中间水平。对于核电而言，全生命周期分析中能源消耗所引起的温室气体排放占其总量的 84%，这意味着核电的温室气体排放水平取决于中国的能源结构。若利用核电及可再生能源代替煤电提供一次能源，则每生产 1 kWh 电力能够减排 1 kg $CO_2$，具有巨大的温室气体减排潜力。2020 年到 2050 年，预计中国会将电源结构中非燃煤发电所占份额从 28%提升至 47%，煤电发电份额由 69%降至 49%。

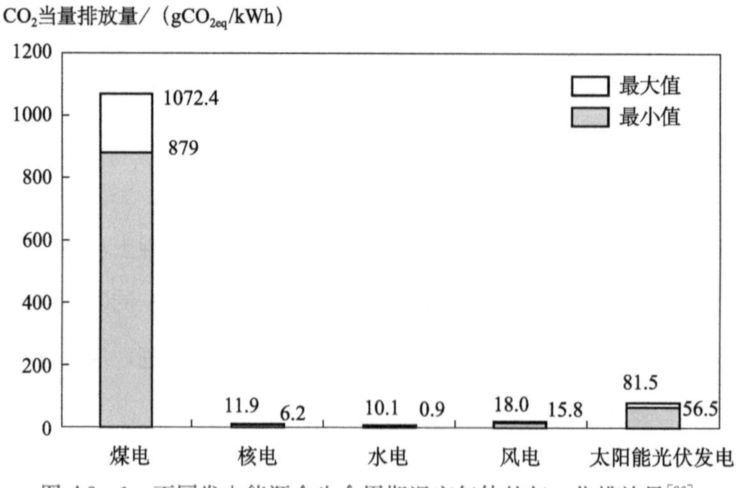


图 A2-1　不同发电能源全生命周期温室气体的归一化排放量[26]

此外，在发电量增加 70%的情况下，温室气体排放仅增加 23%（按不同发电能源的归一化温室气体排放量的高值计）。总之，清洁能源（核电、可再生能源）具有很大的温室气体减排潜力，是构建低碳能源系统、促进电力生产和消费方式转变的关键。

不同发电能源全生命周期所致公众归一化集体剂量（评价范围为各厂址周围 80 km）见图 A2-2。水电、风电以及太阳能光伏发电所致公众剂量相对较低（但目前暂未研究可再生能源资源开发及发电时直接产生的放射性流出物排放，如水电站发电可能导致水中氡的释放等）。核电所致公众剂量的 86%来自于铀矿开采。放眼未来，随着铀矿开采地浸技术份额的提升，中国核电流出物排放所致公众剂量水平还将进一步降低。煤电是目前中国能源结构中最主要的能源，近年来随着装机结构的优化，300 MWe 及以上功率的机组逐渐成为主流，供电煤耗的降低以及除尘技术的发展，使得煤电（除煤灰渣利用外）产生的公众剂量水平明显降低。但燃煤的能量密度相对较低，会产生大量的煤灰渣。这些灰渣经掺混，在中国主要用于制造住宅的主体墙材，这一利用途径产生的公众归一化集体剂量约为 $2.6\times10^{3}$ 人·Sv/GWa（2003～2010 年平均值），约占煤电全生命周期所致公众归一化集体剂量的 99.9%，比其他途径的总和高约 3 个数量级，并显著高于其他发电能源的辐射水平。研究结果表明，改善能源结构、发展核能与可再生能源，能在很大程度上降低公众的辐射照射。

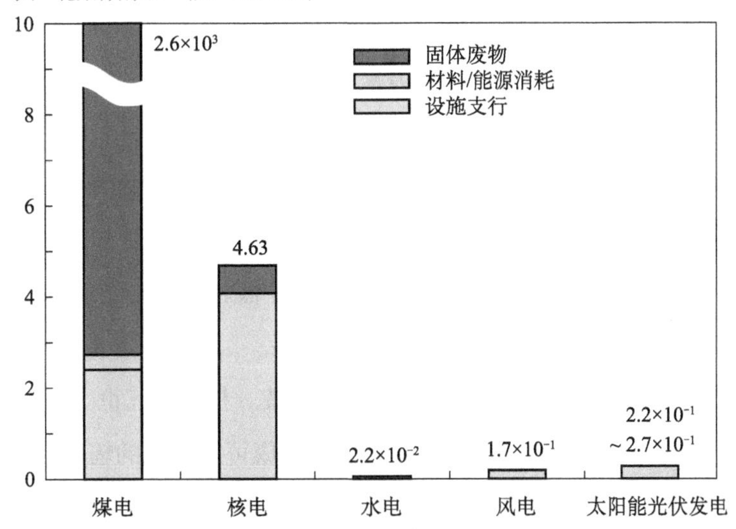


图 A2-2　中国不同发电能源全生命周期所致公众归一化集体剂量

注：“固体废物”在煤电中指煤灰渣在住宅主体墙材上的利用，在核电中指固体废物处置

# 附录 2-2　核厂址周围大型流行病学调查（案例与结果）

释放到核设施周围的放射性物质极少，它们对周围居民生活和健康的影响只能通过流行病研究的方式来评估，这就需要进行概率评价。上述研究分析了由放射性疾病引起的病例或死亡率，以及导致疾病发展的因素。这些研究是在被观测个体的天然环境中进行的，并尽可能地考虑其生活习惯。

第二次世界大战以来，在全世界核厂址及污染区域周围已实施了数百次流行病学调查。

开展流行病学研究并不容易：需要良好的准备工作，准确识别所有干扰参数及其相互关系，对可能影响研究的环境条件进行详细深入的研究，长时间的观察、有序且周密的数据采集、恰当的研究方法和训练有素的专家，以确保正确地解释和处理所获得的数据、所采用的方法以及所得到的结果。

低剂量时，不大容易进行辐射后果分析，原因如下：

（1）发生率低，其影响可能被电离辐射以外的其他原因所掩盖，大多数情况下，电离辐射以单独或同时的方式产生相似的效果。

（2）从方法论和统计学观点出发，由于发生率较低，有必要在几代人中研究非常大的人群样本，以及研究具有类似环境因素且未受到电离辐射照射的非常大的对照人群组（控制样本）。

（3）人类持续地受到天然辐射（如宇宙射线及来自空气和地壳的放射性核素的辐射），以及人工辐射（核辐射的医疗或工业利用）和非电离辐射（如电视和计算机辐射等）。因此，很难区分所造成的影响到底来自何种辐射源。

在这方面，可以列举几项国际流行病学研究：

（1）2006～2010 年，西班牙科学和技术创新部、卡洛斯三世大学健康研究所和核安全委员会共同开展了“核与辐射厂址对人类健康可能造成的辐射影响”研究。研究得出结论：核电厂不会增加人类患癌危险；在所分析的区域内，人类受到的累积剂量评估结果极低，平均值约为核电厂周围天然本底辐射水平的 1/300。

（2）应瑞士医疗当局的要求，瑞士伯尔尼大学于 2008～2010 年开展了相关研究。

结果表明：青少年患癌与瑞士核反应堆厂址之间没有相关性。

（3）法国 IRSN 于 2008 年发布了一份详细报告[27]，对全球核厂址周围的所有白血病（所有类型）的流行病学研究得出综合结论：从局部地区来看，在英国的塞拉菲尔德（Sellafield）和敦雷（Dounreay）后处理设施周围，以及在德国的克鲁梅尔（Kruemmel）核电厂周围，存在着过量的儿童白血病确诊病例。然而，迄今为止，包括法国在内的所有多厂址研究均未显示出核厂址周围年轻人（0～14 岁或 0～24 岁）白血病发病率有所增加。

（4）德国的一项研究显示，德国 16 个核电厂周围 0～4 岁儿童有着过多的小儿白血病案例[28]；但作者提醒读者，鉴于观测到的辐射水平非常低，这一发现有些出乎意料；并指出，儿童白血病的原因仍然无法解释，可能是由于无法控制的干扰因素造成，或纯属巧合。目前，包括法国在内，其他国家开展的研究并不能证实这一观察结果。

许多研究项目试图通过分析多种潜在的危险因素来解释一些核设施周围观察到的过多的白血病例。但是由于对儿童白血病危险因素了解程度不够（特别是胎儿和幼儿期电离辐射照射的潜在影响），致使病因难以确定。不管是在国内还是在国际层面，大规模的调查研究可能是必要的。

一旦核设施建成，通常就会进行流行病学研究。如果没有核设施建造前的情况作为参考，这些研究就无法揭示变化。因此，在核设施建成前和建成后，都需要进行这类研究。

由此看来，为了更好地研究健康效应，在核事故放射性物质释放后对核设施周围进行流行病学研究要比常规调查更为合适，因为常规调查难以揭示设施正常运行时的微小影响。然而，后者作为比较分析的基线，仍然颇具价值。

# 第3章　乏燃料和放射性废物管理

---

**建议**

目前的分析表明，无论是在局部地区还是全球范围，放射性废物管理的现行实践对健康和环境造成的影响都非常小。但是，仍然有必要对导致放射性物质释放到核设施和放射性废物包之外的各种过程加以改进。

中法三院建议，考虑到核燃料循环前端和后端产生的废物以及时间尺度，应当对所有环境影响（辐射和化学影响）及相关风险的评估方法加以改进。

为支持上述总体建议，中法三院提议：

（1）在放射性废物管理过程中的每一步，均应采用最佳可行技术（BAT）来限制/包容放射性核素；

（2）加快放射性废物安全处置进度，确保代际公平，避免给子孙后代带来不当负担；

（3）制订研发计划，更好地了解生态系统受到的放射性和化学影响（可逆性、适应性以及相关元素的生物可用性等），确定用于鉴别放射性废物相关危害的定量参数，从而更好地应对环境问题；

（4）应以公开、透明的方式，采用全面、负责任的体系来保护环境（包括立法体系、主管机构和资金制度等）。

---

## 本章介绍

核工业的特殊性在于其使用的燃料在“燃烧”后不会消失。核工业不能采用化石燃料行业的方式管理废物，即一方面按照标准的废物处置途径以温室气体的形式排放到大气中，另一方面将固体残渣堆积放置。核燃料发生的裂变和其他核反应能够产生大约上百种短寿命和长寿命放射性核素，其中涵盖了元素周期表中 2/3 的元素。所有这些放射性核素的化学性质截然不同。当从核反应堆中卸下时，核燃料的放射性活度从 $kBq/cm^3$（新燃料）上升到 $10^{10}$ 或 $10^{11}$ $Bq/cm^3$（乏燃料）。来自核电厂的所有放射性废物均或多或少含有这些放射性核素。因此，放射性废物管理是核燃料循环的一部分。目前，所有的核能国家都有放射性废物管理的工业渠道。绝大多数放射性废物（放射性水平较低、数量较大）最终都在地表/地表下处置库中处置，剩余的放射性废物（放射性水平较高、数量较少）则在贮存库中贮存，以待深地质处置库的建成。尽管从事此类操作的人员已高度小心，但可裂变材料仍然存在于乏燃料（基本上是钚）中，核废料造成少数放射性核素直接释放到环境中。在遥远的未来（几百年到几十万年），预计一些放射性核素将从处置于岩石圈的放射性废物中返回到生物圈。然而，无论对于何种情形，当前均已采取了应对措施，确保不管地域和时间尺度如何，辐射影响均维持在天然放射性水平的波动范围之内。

本章重点探讨放射性废物管理的环境影响。

## 3.1　放射性废物管理的原则、策略和框架

以下策略和原则反映了国际主流意见，但主要借鉴了法国几十年来的经验。

### 3.1.1　放射性废物管理原则

第一个基本原则是代际公平（即当代人不能将其技术决策带来的负担留给后代）。环境是世世代代共有的财产。给子孙后代留下一个清洁的环境是当代人的重要责任，尤其应抑制天然放射性的增加。第二个基本原则是确保每代人都了解国家和国际层面上放射性废物管理实践。国家和国际放射性废物管理机构应负责将可能产生环境影响的放射性废物位置等信息记录尽可能长时间地保存。

为阻止放射性核素向环境释放，所有核设施的营运单位必须在放射性废物管理中

采用最佳可行技术（BAT），以减少放射性废物的产生。这已经是当前普遍采用的做法，如下所述。

### 3.1.2 放射性废物管理策略

放射性废物管理策略包括：①在反应堆内最大限度地燃烧放射性物质；②浓集和包容放射性核素和有毒物质；③最后在处置库中处置最终的放射性废物。这些专门设计的基础设施旨在尽可能延长放射性核素返回到生物圈的时间（数百年或数十万年），从而将放射性废物与生物圈隔离。同时，避免采用稀释策略。因此，放射性废物管理在本质上与传统废物的管理不同，需要较高的科学技术能力以及强有力的支持。放射性废物管理也需秉承“安全第一”的原则，即使有损经济。

除了如铀矿开采/选矿中产生的超量放射性废物或极低活度水平的放射性废物外，对于燃料循环各阶段产生的原生废物，应尽快处理以进行放射性核素的包容/隔离，或者贮存于设施内，防止其与公众和环境发生任何接触，并留待进一步处置。其目标是生产初级废物包，在包括最终处置在内的所有步骤中进行处理。用于高放废物的包装需要采用最佳技术设计理念，从而对裂变产物、次锕系元素（对乏燃料进行后处理）或乏燃料组件（不对乏燃料进行后处理）（见 2.1 节）进行整备。放射性核素无法从这些包装中逃出（除非在高度假设的情况下）。对于其他的低放废物（放射性要低几个数量级），则无需进行包装密封。

若暂时没有可用的处置库，可将任何类型的放射性废物包贮存在专门设计的设施中，以待最终处置。对于高放射性水平的废物包，贮存设施必须能够使废物包在处置前释放热能，即贮存设施的建设需考虑高放射性水平废物的释热影响。

对环境的短期/中期影响来源于放射性废物的处理和包装。在原生废物处理和废物贮存设施正常运行的情况下，所有释放到环境中的放射性均低于批准限值。对环境的长期影响来源于已处置的废物包，因为容器腐蚀，放射性核素/有毒物质会缓慢释放。在开采铀矿和处置铀矿加工产生的未包装放射性废物的过程中，必须同时考虑这两类影响（短期和长期）。

根据放射性废物的特点，可通过多种方式处置放射性废物，包括地表、近地表和深层地质处置库。不管采用何种处置方式，若要开发一个处置库（其也是一种核设施），就必须进行安全全过程系统分析，以保护公众健康和避免环境污染。此安全全过程系统分析应同时考虑短期/中期和长期影响。

在任何情况下，对人类和环境的短期/中期有害影响必须远远低于法律规定限值且符合现行法规要求。

应根据当前最新信息、数据和景象，利用模拟来评估潜在的长期影响。放射性气体或液态流出物的环境排放会导致放射性扩散，并使其沉积在距离排放源远近不一的位置。最后，放射性核素进入生物地球化学循环。这些排放物的放射性和化学毒性直接影响可以通过就地测量计算出来。但是，在计算放射性核素从受污染土壤、水体或从地质处置的放射性废物包返回生物圈的长期、超长期影响时，则并非这般容易。科学家必须先对放射性核素通过许多天然或外源物质从排放地到岩层出口的时空迁移过程进行模拟，然后才能通过各类景象来评估影响。安全全过程系统分析中需要考虑的时间跨度极大，从一万年（剂量定量评估）到一百万年（定性评估）不等，远远超出了技术领域和稳定社会的一般理论和实践考虑范围。

在实施长期模拟时，需要了解微观及宏观对流或扩散现象的复杂性。对流现象发生在断裂的工程/地质材料中。溶质在无连通裂隙的大型均质材料中的运移依赖于浓度梯度，从而导致理论上的 $t^{1/2}$ 扩散规律。采用的抗降解材料（如高水平/长寿命废物包装）的行为规律，归根结底是微观扩散现象的结果。总的来说，长期模型包含了与时间相关的低功耗经验定律：$t^n$，若 $n<1$，意味着结果随着时间的推移而缩减。支撑这些模型的科学包括地球科学、材料科学和生命科学等领域。每个领域均已经积累了许多数据。

通过地球科学可以了解地质层长达一百万年的热、水文地质、机械和化学（thermal，hydrogeological，mechanical and chemical，THMC）演化机制，并可获得地震、水文地质和气候的区域模型。影响气候的天文周期变化可被纳入全球气候模型。在这些基础上，可在模型中纳入未来几十万年的事件及其后果。

根据自然或人为类比/实验中获得的所有知识，材料科学可以模拟岩石、废物包装体的固体成分和存储结构的演变，对于元素迁移也同样适用。地球化学具有至关重要的意义。因此，在深层围岩中的还原环境（$E_h \ll 0$）使得锕系元素的溶解度及其潜在迁移行为受到很大限制，许多裂变产物也是如此。唯一不易受氧化还原条件影响的长寿命放射性元素主要是 $^{36}Cl$ 或 $^{129}I$。在正常和退化景象下，这些元素在很大程度上主导了预期的排放清单，除此之外的大部分放射性核素仍被阻滞在处置库的近场部分。

模型的建立基础不仅包括几十年来进行的实验，还纳入了旨在加速环境演化的严酷、温和、反复的干扰结果。所获得的结果一般可以在适当尺度上进行机制识别。模型可以通过模拟和将信息进行初始屏蔽的盲验证进行测试。可以通过采用重复实验的方式降低不确定性。外推法在不同领域的极限是已知的。

对于材料科学，可在良好的数学模型基础上建立时间序列模型。对于生命科学，由于新陈代谢的复杂性，时间序列建模并不容易。对于人文科学，历史学家已工作了几个世纪，而社会学家仅有几十年的经验，在形式和逻辑的使用方面并不系统化，因此时间序列模型不够先进，或者根本尚未开展。

在考虑了所有长期模拟的情况下，未来环境影响存在一定不确定性，主要是因为需要对考虑的景象做许多假设。然而，所有的模拟，即使是与最坏的景象相对应的模拟，都表明辐射影响远低于天然放射性的影响（或远低于人为闯入处置库情况下的放射性水平）。

### 3.1.3 放射性废物管理框架

国际放射性废物管理框架如下：

（1）《乏燃料管理安全和放射性废物管理安全联合公约》（IAEA，INFCIRC/546，1997年12月24日）（以下简称《联合公约》）。本公约是1994年至1997年《核安全公约》（IAEA，INFCIRC/449，1994年7月15日）广泛讨论的结果。该公约包含保护环境免受电离辐射影响的章节。公约要求缔约国定期提交报告，说明如何履行《联合公约》阐述的义务。当前，有43个国家向IAEA每三年提交一次报告。

（2）所有国家均需考虑的ICRP建议（IAEA，安全系列115-I，维也纳，1994年）。

欧盟国家还须遵守欧盟议会法令2011/70/Euratom（欧洲原子能共同体）。该法令要求每个成员国制定保护人类与环境的放射性废物管理政策。

此外，IAEA（以及欧盟）制定了一系列有关放射性废物包运输的国际规定，从保护人员和环境角度对容器的坚稳性提出要求。

履行《联合公约》的承诺十分关键。这就是IAEA为何会积极跟进缔约国近三年来的最新情况。

中国和法国都是《联合公约》的缔约国，两国均建立了相对完善的法律和组织框架体系，对乏燃料和放射性废物实施了有效管理。

法国宪法承认预防原则。这一原则适用于所有活动，最终目标是保护环境。其核心是，一旦存在严重和不可逆的环境影响假设，应采取相应措施。这一原则号召开展新的研究，从而更好地理解恐惧现象（2004 年《环境宪章》第 5 节，已纳入法国宪法）。

法国制定了两项放射性废物管理法律：1991 年的法律（侧重于研究）和 2016 年的法律（侧重于决策的执行）。生态转型与团结部（The Ministry of Ecology Transition and Solidarity）负责制定政策并执行政府决策。多个核营运单位参与其中：EDF 运营 58 个核反应堆，奥兰诺循环（Orano-Cycle）和法马通公司运营核燃料循环前端和后端设施，CEA 引领核能研发，法国放射性废物管理局（Agence National pour la Gestion des Déchets Radioactifs，Andra）负责放射性废物的长期管理。法国核安全局（Autorité de Sûreté Nucléaire，ASN）代表法国政府确保对核安全和辐射防护进行控制，以保护人民和环境免受核活动带来的风险。IRSN 是 ASN 在安全全过程系统分析方面的技术支持单位，并在放射性废物毒性和放射性废物环境影响方面自行开展研究。最后，还成立了一个核安全透明度和信息高级委员会。该委员会负责制定并提出相关建议，以提高公众信息的透明度和质量。

在法国，所有与放射性废物管理有关的研究和调查都被编入《国家放射性材料与废物管理计划》（*Plan National de Gestion des Matières et Déchets Radioactifs*，PNGMDR）。该计划由 ASN、生态转型与团结部和多个工作组（包括环境保护团体）每三年发布一次，是执行废物管理政策和进行公众宣传的战略工具。该计划能够确定需求、设定目标、向所有利益相关方提出建议，并强制要求提交报告。该计划由一个负责环境的委员会负责评估，之后提交给议会审议并征求公众意见。

Andra（国家放射性废物管理机构）每三年收集一次放射性废物产生单位的申报信息，以此为基础发布一份放射性废物和核材料的完整清单。清单包括放射性废物的种类、数量、当时和未来几十年的地理位置，这些信息以国内核设施的多个演化情景为基础。任何情景变化都会对废物的性质和数量产生影响，并可能对环境造成影响。

此外，还依法设立了一个独立的国家评估委员会（Commission Nationale d'Evaluation，CNE），每年负责向议会报告这一主题的研究结果。

中国建立并实施了一套乏燃料和放射性废物管理法律框架，包括相关国家法律、行政法规、部门规章、管理导则和参考性文件，以及乏燃料和放射性废物管理活动许可制度。相关国家法律包括2003年全国人大常委会（National People's Congress Standing Committee，NPCSC）颁布实施的《放射性污染防治法》（*Law on Prevention and Control of Radioactive Pollution*，LPCRP）和2017年全国人大常委会颁布实施的《核安全法》。

中国生态环境部/国家核安全局（Ministry of Ecology and Environment/National Nuclear Safety Administration，MEE/NNSA）负责乏燃料和放射性废物的监管，中国国家原子能机构（Chinese Atomic Energy Authority，CAEA）是乏燃料和放射性废物管理的主管机构。

在中国，放射性废物产生单位承担放射性废物管理的全面安全责任，并对放射性废物实施分类管理。

## 3.2 放射性废物的特性和分类

有核国家会根据本国工业状况调整放射性废物的分类，因而各国对放射性废物类别的管理实践可能有所不同，但也有许多共同之处。中法三院根据放射性废物中放射性核素的活度和寿命，在第一次报告中讨论了这一主题。后续使用的名称如下：极低放废物（very low-level waste，VLLW）、长寿命低放废物（low level-long lived waste，LL-LLW）、短寿命低中放废物（low and intermediate level-short lived waste，LIL-SLW）、长寿命中放废物（intermediate level-long lived waste，IL-LLW）和高放废物（high level waste，HLW）。

必须指出，放射性水平最高的放射性废物（IL-LLW和HLW）的类型取决于有核国家就乏燃料循环所做出的决策。

### 3.2.1 乏燃料和/或后处理放射性废物

核反应堆产生的放射性废物的管理工作涉及：

（1）乏燃料（乏燃料作为放射性废物）和乏燃料后处理产生的放射性废物（乏燃

料作为裂变材料的来源)。

(2) 核反应堆或核燃料循环设施产生的所有其他放射性废物。

正如前文所述，存在两种类型的核燃料循环。

● 开式核燃料循环：在此循环下，乏燃料不进行后处理而进行湿法或干法贮存，待充分整备且作为 HLW 封装后转移以待最终处置，待处置乏燃料废物包经久耐用，可达数千年。

● 闭式核燃料循环：在此循环下，对乏燃料进行后处理、提取有价值的材料（U 和 Pu)、剩余材料以玻璃固化体形式进行整备并作为 HLW 和 IL-LLW 封装。待处置的玻璃化废物包装经久耐用，可达数千年。

乏燃料和玻璃固化体的放射性大致相同，但后者不含钚和铀。

根据本国的政治、经济、技术或外交背景，全球所有有核国家会采用上述任一核燃料循环方式。基于可持续的核能政策，法国和中国选择了闭式核燃料循环方案。欧盟 28 国（包括英国）选择的循环方式不尽相同。一些国家的选择方案只是暂行的，可能会随着技术进步以及更先进的有核国家在可裂变放射性核素回收方面的经验提升而发生变化，还有一些国家并没有核电设施。

根据待处置的乏燃料和核玻璃固化体的包装特点，预计两者的长期环境影响相差无几。尽管乏燃料和核玻璃固化体的贮存条件不同（乏燃料贮存在水池或干桶中，玻璃固化体则贮存在干设施中)，但当处于监控情况下时，两者的长期环境影响也无明显不同。相比之下，后处理（将放射性核素从乏燃料中化学分离）对环境的区域影响要高于不进行后处理的情况。

简单起见，本章将不涉及核燃料循环问题，而主要探讨放射性废物管理的主要趋势，以更好地论述可能造成环境影响的现象。

### 3.2.2　放射性废物的特性

与化石能源相比，核能每 MWh 产生的废物要少得多。这是因为核能的能量密度非常高，约为化石燃料的数千倍，具体取决于核燃料的燃耗和反应堆类型。与产生气态和固态残渣的化石燃料火电厂相比，产生放射性废物的核电厂和核燃料循环设施的数量要少得多。因此，安全和环保当局更容易按照被有核国家接受且符合国际机构建议的严格规程进行放射性废物管理。应当对所有放射性废物流的流转过程和特点进行充分、详实的记录。有核国家对早期核电反应堆和军事遗留的放射性废物均进行了清楚无误的库存记录。

放射性废物管理的主要环境影响包括，使公众暴露于电离辐射环境下，以及改变水生和陆地生态系统（可能导致生物多样性丧失）。放射性废物管理的环境影响主要是严重事故下放射性物质的液态或气态释放导致的；其他影响还包括，导致那些严重依赖重要原材料持续输送的设施中断作业（交通拥挤、噪声……）。

可按照试验方法对含有放射性或有毒物质的气体或液体排放产生的剂量（外部和内部）进行评估，其结果将被提交给国际机构审查（全球循环比对试验）。由此，就能更好地理解各过程中涉及的现象，并使用开放数据库来提供模拟模型。同样，也可参照世界卫生组织（WHO）的建议，采用类似方法来评估有毒化学品的影响。

由于仍然缺乏非人类生物圈的数据，因此很难量化放射性和毒性对其他生态系统的影响。与其他物种相比，人类通常对辐射危害更为敏感。当人类获得充分的辐射防护时，非人类物种也将得到适当的保护。

在这两种情况下，主要知识的缺口都与放射性元素的生物有效性有关，即与可转移到生物体内的放射性化学物质有关。就这一点来说，元素的物种形成/种态极为重要。环境中的放射性核素浓度处于示踪剂水平，其物理化学行为无法从其所属元素在常规浓度下的行为来推断。放射性核素的种态不控制化学系统，相反，其在可计量的数量上受到该种态施加的限制。放射性元素的化学特性丧失了。原则上，放射性单体种态可能存在，但它们通常吸附在天然胶体上，表现为放射性假胶体。相比之下，废物包中放射性核素的行为是对应元素在常规浓度下的行为。需要强调的是，溶解度现象限制了核素的释放。为弄清放射性向生物体转移过程中所涉及的基本现象，目前已经制定了许多国家和国际研究方案。这一专题应列入到旨在了解人类对环境影响的跨学科调查之中。放射性核素的生态毒理学研究虽已进行了几十年，但进展缓慢。

其他能源工业也会产生各类包含有毒有害物质的非放射性废物。为此，可能需要在垃圾填埋场进行特殊处理和/或处置；但是，尚未产生毒性作用的化学品浓度要远远高于尚未造成放射性危害的放射性核素浓度，因此封隔化学品的传统处置场可能不及封隔放射性废物的处置库那般有效。显然，用于封隔放射性核素的放射性废物处置库也同样能够有效封隔有毒物质。

### 3.2.3　放射性废物的分类

通常情况下，核电厂会定期卸出乏燃料并以新燃料取代。正如之前指出的，会对乏燃料进行后处理或作为高放废物贮存和处置。若进行后处理，乏燃料棒将被切开，以分离回收钚和铀。目前，后处理产生的所有其他放射性核素均溶解在玻璃体中并被封装。封装的废物与处理前的乏燃料一样具有放射性，但钚和铀的含量非常低。后处理和制造过程会产生额外的放射性废物流，这些废物流或多或少会被乏燃料中的长寿命放射性核素所污染。同样，核电厂运行也会产生放射性废物。最后，预计核反应堆和设施在未来的大规模退役将产生各类大量的放射性废物，但主要是低放废物。

就环境影响而言，区分短寿命和长寿命放射性废物具有重要意义。事实上，短寿命放射性废物通常在地表/近地表设施中处置，因此，在逻辑上，其显然会对当代环境造成直接影响。长寿命放射性废物通常在数百米以下的深地质处置库中处置，因此，环境影响很可能在遥远的未来才会发生。尽管如此，对这两种策略都需仔细研究。

此外，还应区别放射性废物中的放射性核素来源是天然（U、Th和子体）还是人造（锕系元素、裂变产物、氚等）的。仅含有铀的放射性废物来源于核燃料循环前端。与核燃料循环后端有关的放射性废物还含有许多其他放射性核素。

放射性核素或有毒物质的释放可能会发生在设施/处置库运行期间，甚至是设施/处置库关闭后。放置废物包装的操作程序和设备，以及日常和定期的放射性（和化学污染物）监控都可检测到影响正常运行的故障。因此，须根据可能发生环境影响的地域来评估当地环境影响。在处置库关闭后的很长一段时间，放射性核素和有毒物质的释放概率及程度取决于废物包装的坚稳性，以及工程屏障和天然屏障阻滞元素向生物圈迁移的能力。如前所述，只能通过模拟方式评估近场和远场影响。在法国，放射性废物的分类遵循开式核燃料循环方案中的常规方法。目前，已经建立了所有放射性废物的处置路径，但LL-LLW、IL-LLW和HLW除外（见附录3）。对于LL-LLW，Andra正在进行地质调查，以期在黏土层建设地下处置库。对于IL-LLW和HLW，Andra计划在2019年前向ASN申请Cigeo（Centre Industriel de Stockage Géologique）处置库场址许可。Cigeo是一个位于500 m深、130 m厚的黏土层深地质处置库，旨在接收所有不能在地下处置库中处置的放射性废物。需处置的放射性废物包括目前核电厂和核燃料循环当中所有反应堆和设施所产生和将要产生的废物，无论政府今后将作出何种能源选择。政府目前的选择是，从 $UO_x$ 乏燃料中一次性回收钚和铀，并将 MOX 乏燃料贮存，作为启动快中子反应堆的有价值核材料。此处的假设是，所有 $UO_x$ 乏燃料都将进行后处理。

从放射性物质中分离出放射性废物和有用核材料是产生单位的责任。

中国依照 IAEA 放射性废物分类安全标准建立放射性分类体系，现行的分类体系等效采用了 IAEA 于 2009 年发布的《放射性废物分类》安全标准（GSG-1（GSG，General Safety Guide 为通用安全指南））[29]。中国的放射性废物分类体系与法国类似，都是基于处置策略分类。主要的区别是，中国的低放废物对应于法国的短寿命低中放废物和低活度长寿命低放废物，而中放废物对应于法国的高活度长寿命低放废物。

## 3.3 放射性废物处理和流出物排放

### 3.3.1 放射性废物最小化

当待处理的原生放射性废物量最低时，放射性废物管理导致环境影响的风险最小。要使放射性废物的量最小化，首先要对所有设施产生的放射性物质进行分类，如此可以排除处于放射性检测极限或低于清洁解控水平的废物（如果存在）。下一步是对放射性废物进行包装，以减少放射性核素在运输和贮存过程中的扩散。目前存在许多包装技术，因此可在包装和贮存造成的直接环境影响和地质处置造成的间接环境影响之间找到最经济的方案。但在任何情况下，通常都需要采用最佳可行技术(BAT)。

几乎所有国家都设定了放射性废物的清洁解控水平或检测限值，这使得非放射性材料中潜在放射性废物的分类被取消。这种可用于公共用途的材料的解控使得最丰富的 VLLW 的数量大大减少。这涉及放射性物质的豁免概念和清洁解控概念。第一个概念用于对有限数量物质（例如 1 t）的放射性浓度（Bq/g 或 $Bq/cm^2$ 或总活度）进行判定，低于该限值，则无需进行任何控制来确保辐射防护，或当使用回收材料时，环境影响可忽略不计。第二个概念需要考虑去污材料再利用时的放射性浓度（Bq/g 或 Bq/cm$^2$ 或总活度小于/等于豁免值）。通用清洁解控水平应确保在任何不利情景下，辐射影响每年低于 0.01 mSv（IAEA 安全准则 RS-G-1.7 和欧洲原子能共同体指令 96/29 的建议剂量）。如此低的剂量不会对环境产生影响。

另一种减少放射性废物量的方法是回收类似低放废物的金属材料。它们可以使用去污方式进行熔炼。熔炼是唯一一个可使回收材料的放射性实现均匀化的过程，也便于放射性监测。

似乎不大可能通过再循环来减少核燃料循环过程中产生的其他放射性废物的量。

法国不对 VLLW 进行豁免和解控。法国核安全局（ASN）认为，所有可能接触过放射性污染或被辐射激活的材料都是受管制的 VLLW。ASN 表示，不进行豁免和清洁解控的主要原因是，这些概念很难应用，因此，个人所受的剂量限值为 0.01 mSv/a。ASN 认为，虽然清洁解控有很多优势，但缺点是无法考虑所有可能的情景；他们指出，仍未进行安全分析参数讨论，而且很难在工业规模上实施放射性测量协议；最后还有一个风险，即人工放射性将像天然放射性一样无处不在。法国的立场与国际建议和欧洲的部分实践不一致，因此正在接受审议。尽管如此，对于在固体材料（与人类接触的产品除外）中添加放射性核素的特殊情况，可以给予一些例外的解控许可（有条件的清洁解控）。此外，在核工业中亦可对只受极低放射性污染且可监测的特殊材料进行回收。此类案例将提交给 ASN 进行特定批准（一事一议）。

几年来，废物产生单位、Andra、IRSN 和 ASN 一直在研究非常、非常低水平的废物（very，very low-level waste，VVLLW）的解控阈值设定。EDF 和 Orano 正在研究相关的技术和经济条件，从而能够通过熔炼大量的金属放射性废料进行回收利用。

中国已开始实施关于放射性废物最小化的核安全导则。新建核电厂均符合该导则的要求。运行中的核电厂已采取切实可行的措施，落实放射性废物最小化的原则。

### 3.3.2　流出物排放

如前所述，向环境中释放的气态或液态流出物是环境所受直接影响的主要来源。对于放射性废物管理而言，流出物与初始放射性废物包同样重要。必要时，可对气态流出物进行过滤或利用适宜的溶液冲洗，以便去污。这将产生二次固体废物和净化后的气体。这些气体将根据监管要求排放到大气中。

在对核设施工艺中产生的液态流出物进行局部处理后，会产生净化的液体溶液，并按照授权规定排放到环境中；处理后，还会产生固体放射性废物和放射性浓缩液体（将转化为固体形式）。

在中国，放射性废气和废液需经过处理，使放射性水平尽可能地低，以便满足排放要求。气态和液态流出物排放均需进行监测和控制，确保不发生意外排放情况。应选择适当位置排放液态流出物，液态流出物在受纳水体中得到充分稀释后，应达到近零排放要求。

## 3.4 放射性废物处置

### 3.4.1 极低放废物（VLLW）

对于低活度/极低活度（小于 $10^2$ Bq/g）的放射性废物，甚至是含有微量长寿命放射性核素（如铀）的放射性废物，世界上的大多数有核国家基本上都采用填埋场处置（地表或近地表）。事实上，由于放射性很低，放射性核素的半衰期并不是一个决定性因素。一般来说，这种放射性废物数量很大。与短寿命放射性废物处置方式的不同之处在于，其仅需要简单包装和一个规模相当有限的工程基础设施，且包装无需具备封隔放射性核素或有毒物质的功能。定义短寿命放射性废物的规则也同样适用于这类放射性废物：控制放射性废物（包装、袋子、材料等）、根据预先确定的容量控制设施填充量、控制释放、控制环境。目前，全世界有很多类型的废物处理设施，其能够处理技术放射性废物、仅含天然放射性核素的黄饼加工放射性废物，以及铀浓缩产生的放射性废物，IAEA 建议通过挖沟进行地表/近地表管理。

法国将必须处置超过 20 亿立方米的 VLLW，大约是目前处置库容量的 4～5 倍。VLLW 主要产生于反应堆和核设施的拆除过程。Andra 将对现有处置库进行扩容。Andra、废物产生单位、ASN 和 IRSN 正寻求一种新的管理方案：新建一个中央处置或集中处置中心，循环利用金属废物和混凝土并有条件解控 VVLLW。

如前所述，法国没有解控 VLLW 的实践。

中国拥有 4 座在运的 VLLW 填埋场，至今共处置了 1 万立方米的 VLLW。

### 3.4.2 短寿命低中放废物（LIL-SLW）

短寿命放射性废物（$10^2$～$10^6$ Bq/g）主要产生于核电厂的运行。一些短寿命放射性废物中含有极少量的长寿命放射性核素。短寿命放射性废物包装通常在专门设计的地表/近地表设施中处置，深度可达数十米。包装物可以是密封或不密封的钢桶、混凝土桶或大容器。安全与环保当局负责根据处置场址的特征、废物包结构和工程屏障确定处置库关闭时可接受的放射性容量，以及各放射性核素或有毒材料的容量。地表处置设施容量限值确定的考虑因素是，当数百年后短寿命放射性核素消失（但长寿命放射性核素不会消失）时，放射性废物处置场能够恢复到绿地状态。为此，需对经核准的气体和液体流出物释放进行监管和控制，同时进行环境监测。所有的批准均需符合处置场/设施的安全分析结论要求。

在长达数十年的运行过程中，化学物种从废物包转移到环境中的主要媒介是雨水或地下水流。如有必要，可收集并处理雨水，同时在处置场出口处进行地下水监测。此外，气体也可能会从废物包装中逸出（如以氚化氢和氚化水形式存在的氚），这就是为何要对处置库的氚容量设限。

在对短寿命放射性废物处置库进行选址、运行和监控时，应考虑如何支持新技术的采用，从而改善放射性核素封隔性，减少放射性物质向环境瞬时或长期释放，并作为地表/近地表放射性废物管理工作的参考。

当处置场恢复到绿地状态时，预计其影响将仅限于安全全过程系统分析中的范围。

含有大量氚的废物存放在贮存设施中，以待氚活度降低，然后按照与其分类相对应的工业渠道进行管理。

法国将必须处置现有核领域产生的 LIL-SLW，数量达（15～20）亿立方米。法国 25 年的 LIL-SLW 处置库（50 万立方米，20 多年前关闭）运营经验表明，虽然氚很难封隔，但对公众的影响每年不到零点几微西弗（μSv/a）（见附录 3）。

中国低放废物中主要含有短寿命放射性核素，而长寿命放射性核素数量有限。这类废物可在近地表设施中处置，其相当于法国的低、中水平短寿命废物。中国已在两个近地表处置设施[30] 中处置了约 2 万立方米的放射性废物。

### 3.4.3 长寿命低放废物（LL-LLW）

由于低水平长寿命放射性废物（10～$10^5$Bq/g）含有的一些放射性核素（如$^{35}Cl$或$^{14}C$）难以通过工程屏障或天然屏障封隔，且数量大到无法在深层地质处置库中处置，因而不可在 LIL-SLW 或 LLW 处置库中处置该类放射性废物。如果考虑近地表处置，则必须在选择处置场址时考虑这类放射性核素的长时间封隔要求。因此，处置库必须足够深，以保证天然屏障的厚度足够并发挥良好功能。

预计现有核领域的 LL-LLW 总量为 19 万立方米。Andra 将继续根据两个地下处置概念来确定潜在的黏土场址。初步概念应该在几年后完成。如果缺乏处置库，会导致放射性废物长期贮存，并减缓拆卸速度。

中国现行的废物分类体系中没有低水平长寿命废物的分类。对于主要含长寿命核素的放射性废物（10～$10^5$Bq/g），核素活度浓度低于低放废物上限值的废物属于低放废物，可在近地表设施中处置；核素活度浓度高于低放废物活度浓度上限值的属于中放废物，适用于中等深度处置。

### 3.4.4 长寿命中放废物（IL-LLW）和高放废物（HLW）

核专家表示，在采用多重屏障设计的情况下，可在深地质结构中进行 IL-LLW（$10^6$～$10^9$Bq/g）和 HLW（$10^9$Bq/g 及以上）的环境隔离和放射性核素封隔。此地层必须已经保持稳定数亿年，且具有良好的地球化学性质，如限制水循环和滞留化学元素。选择对高放废物进行地质处置的根本原因，是出于社会学中的社会稳定性考虑，而这种社会稳定性是几个世纪以来都无法保证的。因此，更合理的做法是，依托地质学知识，在相当长的一段时间内使这类废物远离生物圈。

无论选择哪种核燃料循环模式，在经过较长的临时存储期（例如，冷却水池存储或干燥存储）后，放射性废物的热辐射均会减少，因而可在深地层中处置 HLW 和 IL-LLW。为此，必须设计特殊地面设施以容纳这些废物包，以待进一步处置。初始废物包在进行处置前，将置于外包装中。在贮存期间和将废物包放入到处置库期间产生的环境影响，与核设施正常或意外操作期间的环境影响相同，尤其是在批准排放的情况下。

深地质处置库用于接收所有在地表/近地表处置库中无法处置的放射性废物。在减少放射性核素活度方面，不对深层处置库的容量作出限制。

可根据深地质处置库场址的地质岩层（例如黏土或花岗岩）确定采用不同的深地质处置库概念。黏土能够减缓并最终阻滞乏燃料中所有放射性核素的迁移，由于黏土容量极高，可通过各类机制捕获放射性核素。这正是在选择花岗岩作为岩层时要使用黏土作为缓冲回填材料的原因。为此，各国已进行了广泛而深入的调查，目前仍在寻找合适的深地质处置库场址。

到目前为止，仅芬兰在花岗岩中成功开挖竖井，其目的是建造一个深达450 m的地下乏核燃料处置库（名为“安克罗（Onkalo）”）。瑞典也即将完成地质处置项目。乏燃料将封装在铜容器中，并以膨润土环环绕的方式贮存在斯堪的纳维亚地盾花岗岩的钻孔中（KBS-3概念）。在密封处置库之前，所有通道和竖井都将使用膨润土填充。法国准备申请一个许可，以允许几年后在500 m深的黏土中建造一个处置后处理产生的IL-LLW和HLW的处置库。在法国，装有玻璃固化体的外包装将放置在水平隧道中，IL-LLW的外包装将放置在130 m厚的水平延伸黏土层（Callovo-Oxfordian黏土）中心竖向开挖的大洞室中。所有的工程结构以及通道和竖井都将使用特殊混凝土/膨润土塞密封。此类工程结构预期可封隔放射性核素（和有毒物质）相当长的一段时期（几千年到几万年），从而避免对生物圈产生任何影响。

在过去数十年间，有十个国家在积极准备，希望在未来数十年内启用处置库。由于流程较多（如场址表征、分析和最终选择），涉及大量科研工作，以及政客和公众亦参与了决策过程，地质处置库的实施工作历时弥久。这些国家都准备了许多有关地下处置库选址的国家计划报告。国际组织（欧盟、经合组织核能机构、国际原子能机构）也为此设立了国际联合研究项目，旨在了解控制放射性核素迁移的基本现象并测试工程屏障。

按照计划（如法国计划），深地质处置库将持续运行一个半世纪之久。在这段时间里，尽管已采取了防范措施，但仍然可能因意外情况产生一定的环境影响。可通过模拟并与当前情况对比来评估环境影响。同时，在启用处置库之前，应当对环境进行长时间监测。

对处置库关闭后各组分长期演化进行模拟，是安全全过程系统分析的主要关注点。尽管 IL-LLW 和 HLW 废物包装对放射性核素的隔离和包容能力很强，但也会被逐渐腐蚀，随之释放出放射性废物。放射性核素和其他元素的迁移将开始缓慢进行。根据大量浸出实验和天然类似物调查，玻璃体或铀氧化物包装的寿命估计将超过几十万年。大量模拟结果表明，锕系元素在黏土中的迁移距离不会超过 10 m，而可移动裂变产物在到达生物圈时的时间很长，放射性活度已急剧下降。

通过模拟放射性核素向环境中的迁移，可以计算处置库场址出口处的长寿命放射性核素浓度。然后，根据土地和水的使用情景，采用目前使用的标准方法推断个人剂量。

对长达数百万年的行为进行多次模拟的结果表明，放射性废物包装、深地质处置库的天然和工程屏障非常有效，处置库中放射性核素的释放所导致的剂量将不超过天然本底放射性剂量的 1/10（见 3.1.2 节）。

在为获得地质处置库建造许可而向核安全和环境当局提交的文件中，需包含所有的数据、实验和模拟结果，以及对处置库的安全全过程系统分析。

根据目前的核燃料循环策略，预计需要在法国深地质处置库中处置约 72000 $m^3$ 的 IL-LLW 和 12000 $m^3$ 的 HLW。这些废物正在贮存中，以待 Cigeo 试运行成功。

在中国现行的放射性废物分类体系中，中放废物含有相当数量的长寿命放射性核素，因此需要采取高于近地表处置包容和隔离水平的措施，使之远离生物圈。中放废物处置深度通常在几十到几百米。如果处置系统选择了合适的天然屏障和工程屏障，则该深度的处置有可能实现长期的环境隔离。特别是，这种深度的浸出一般不会在短期和中期产生有害影响。与低放废物近地表处置相比，中等深度处置的另外一个重要优点是，可极大降低人类意外闯入的概率。因此，中等深度处置设施的长期安全无需依赖有组织控制。

中国计划 2050 年左右建成深地质处置库[31]。目前，中国制定了高放废物处置总体研发规划。确定甘肃北山为我国高放废物处置首选预选区。已开展高放废物地质处置缓冲回填材料研发，目前正在开展放射性核素迁移和安全评价研究。地下实验室（underground research laboratory，URL）场址和建设方案已经确定，2021 年开工建设，计划于 2027 年建成中国首个地下实验室——北山地下实验室。

### 3.4.5　铀燃料循环（铀矿开采）前端产生的仅含有天然放射性核素的放射性废物

铀矿开采产生的大量铀放射性废物由尾矿、矿石加工（为获取黄饼）废物残渣以及其他技术废料组成。该放射性废物包含铀、铀的所有不挥发子体以及其他化学物质（$^{226}Ra$ 是唯一一种含量较大的化学物质）。

在黄饼精炼及其转化为 $^{235}U$ 富集的气体氟化物过程中，会产生大量只含有天然放射性核素的放射性废物。

1. 采矿

法国已有 50 年的铀矿开采历史，目前已停止铀矿开采作业。在此期间，已从 250 个场地开采了 5200 万吨矿石，并提取了 8 万吨铀。采矿的放射性废物来源于 1.66 亿吨的开挖岩石，包括尾矿和 5200 万吨的采矿残渣。与矿场材料和残渣有关的铀含量数量级和放射性水平如表 3 - 1 所示[20]。

表 3 - 1　铀矿开采活动和天然岩层中的铀含量和放射性水平

| | 铀含量/（g/t） | $^{226}Ra$ 活度浓度/（Bq/kg） | 总活度浓度/（Bq/kg） |
|---|---|---|---|
| 法国土壤和岩石的平均数 | 几个 | 几十 | 几百 |
| 花岗岩 | 几十 | 几百 | 几千 |
| 矿石 | 约一千 | 几万 | 几万 |
| 废料 | 几十到几百 | 几十万 | 几万到几十万 |
| 残渣 | 几百 | 几万 | 几十万 |

采矿废料在大面积开挖后被就地处置，且需在封闭处覆盖一层天然材料，以防止氡释放和直接射线照射。对于采矿废料的监控，需重点关注设施出口处雨水和地下水中的氡释放、铀和有毒物质释放。由于水在排放到环境中之前经过处理和净化，因此仅含少量铀和镭。同时，也需要监测这些放射性核素在环境中的积聚情况，并在必要时定期采取补救措施。就各类铀燃料循环步骤而言，开采、选矿和矿石浸出是污染最严重的步骤。

在法国，残渣和尾矿贮存在 17 个 ICPE 处置库中。这类残渣（黏质砂土或

$H_2SO_4$ 浸出的矿石块）放置于土工聚合物基底并使用 2 m 厚的废料和 0.4 m 厚的土壤覆盖。对于渗滤水（6000 $m^3/a$），既可在排放前处理，也可在回收部分铀之后处理。对残渣中岩心样品的监测和分析反馈表明，经水和成岩作用侵蚀后的残渣具有一定稳定性。此外，U 和 Ra 能够被一些矿物相捕获。铀吸附在黏土矿物和铁（Ⅲ）的氢氧化合物上，并形成不溶性 U（Ⅵ）/U（Ⅳ）复合磷酸盐。镭与 $BaSO_4$ 共沉淀后，也吸附在黏土矿物上。目前，已经建立了 U 和 Ra 的行为模型。总的来说，U 和 Ra 的流动性不强。从环境监测和现场定期分析中获得的所有信息都可用于这些元素的迁移建模。

法国国内（和尼日尔）经验表明，仍然需要注意尾矿和采矿废料可能被重新用作建筑或石渣料（尼日尔也会再次利用受污染的废铁）这个重大问题。在法国，当局部影响大于 0.6 mSv/a 时，就需要恢复采用废尾矿扩散的辐射标准。所有的尾矿和采矿废石目前都存储在处置库中。对于在尾矿和采矿废石上建造的建筑物，其所致个人剂量达到 0.5～1 mSv/a，氡浓度超过 1000 $Bq/m^3$ 时，影响最大。而根据法国卫生条例，公众每年的个人剂量不得超过 1 mSv，氡浓度不超过 300 $Bq/m^3$。由于经常使用尾矿压载的区域仍然存在，但引入的剂量明显较低（至少一个数量级），因此不成问题。

中国目前有 80 个铀矿，其中 30 多个已退役。采矿放射性废物包括 3400 万吨的挖掘岩石和尾矿以及 1100 万吨采矿废渣[32]。

法国的铀浓缩由奥拉诺公司（Orano）进行。在现场对放射性废物进行管理。

**2. 炼制黄饼产生的放射性废物**

Orano 转换设施位于法国南部的马尔维斯。迄今为止，产生的放射性废物仍留在现场，新硝酸盐废水贮存在大型蒸发池（70000 $m^3$）中，其他废水（约 280000 $m^3$）贮存在近地表沉积物中。固体硝酸盐将从“蒸发池”的硝酸盐废水中收集，并加工成现有废物清单中的 VLLW 和 IL-LLW。一个新的设施最近已投入使用。未来，采用新工艺产生的放射性废物将按照 VLLW 和 LL-LLW 进行管理。

## 3.5 开式/闭式核燃料循环产生的废物

放射性废物管理的环境影响程度，与反应堆和设施运行（包括采矿）所释放的放射性核素和产生的放射性废物量有关。可根据这些指标对两种核燃料循环进行比较。法国CEA[33]对法国实际运行的开式核燃料循环（open fuel cycle，OFC）和单次回收钚（close fuel cycle，CFC）的闭式核燃料循环的估算见2.4.1节。本节提请注意这样一个事实，CFC中的后处理会向大气释放大量的惰性放射性气体和氚（$5.5\times10^{11}$ Bq/TWh），同时向海中排放一些低放射性液体（$2.24\times10^{10}$ Bq/TWh），但无明显放射性影响。对于这两种核燃料循环，LLW和LIL-SLW的产量没有显著差异，但是CFC产生的IL-LLW（1.18 $m^3$/TWh）是OFC（0.32 $m^3$/TWh）的4倍，CFC产生的HLW（0.36 $m^3$/TWh）则是OFC（1.17 $m^3$/TWh）的1/4。

由于处置库运行产生的局部环境影响因设施各异，处置库运营单位负责每年向安全和环境当局报告相关工作。关于放射性影响，估计公众剂量不到1/10 μSv。长期影响的目标是在所有情况下均符合当前的放射性目标，即小于1 mSv/a。

对于CFC而言，需要考虑的问题是，目前的反应堆将被第三代反应堆（通常是欧洲压水堆（European Pressurized Water Reactor，EPR））所取代，产生的放射性废物将减少20%～35%（取决于反应堆类型），这是因为，第三代反应堆在热力学效率、燃耗率和使用寿命方面性能更佳。尽管如此，后处理和反应堆运行过程中的液体释放量将增加20%左右，但与天然辐射相比，辐射影响仍然很小。

## 3.6　新技术

如果采用SFR（或更广泛地说，第四代快中子反应堆发电），由于取消了所有的前端循环操作，废物释放量和生成量将大幅减少。此外，在降低长寿命放射性废物影响方面还能实现另一个重大改进：在后处理阶段，从乏燃料中提取次锕系元素（Am、Cm、Np），并使用第四代快中子反应堆或混合反应堆进行燃烧。理论上讲，这一附加的乏燃料后处理步骤可将HLW危害的持续时间从几十万年缩短至几百年。在法国，已经通过千克单位的小规模试点证明了提取次锕系元素的可行性。但只有当第四代快中子反应堆趋于成熟后，才可能扩大到工业规模。研究表明，只有当所有SFR都能嬗变锕系元素时，才有可能发生次锕系元素的嬗变。这一要求意味着核电生产将发生巨大变化：新的SFR、新的闭式核燃料循环、新的提取方法、新的燃料制备等。就环境而言，进行化学处理的放射性物质越多，放射性核素释放的风险越大。

此外，由于经济原因，次锕系元素的嬗变显然不能应用于已经封装在玻璃体中的材料。

使用ADS进行次锕系元素嬗变的其他方法正在研究之中。最先进的项目是比利时的Myrrha。关于ADS-嬗变对环境的影响，只有一些前瞻性信息。无论ADS的性能如何，都必须准备好嬗变靶，并能够将其回收以获得良好的嬗变率。放射性物质的分离总是会导致有限数量的放射性核素不可避免地释放到环境中。

法国在过去的十年里已制订了一项雄心勃勃的研究计划，准备在2040年左右启动第一台第四代商用SFR，但如今正在重新考虑这一目标。

中国的目标是，在2035年左右推出第一台商用SFR，2050年左右开始大规模建设。

中国已启动用于分离和嬗变的ADS项目，计划于2050年建成示范项目。

## 3.7 结论

在放射性废物管理的每个环节，基本都要从环境保护角度考虑废物包的隔离/包容、贮存，以及针对各类型放射性废物的设施处置。所有的措施都会采用最尖端的技术，且在研究工程屏障和岩石圈中放射性核素/有毒物质行为的过程中，能够得到持续的研发支持。

需要对包括放射性废物产生到处置库处置在内的所有活动进行监测，以确保放射性废物包与生物圈隔离。同时，还要持续监测这些设施周围环境的本底水平。放射性核素的正常释放对环境的影响不得超过安全和环保当局在颁发设施许可时批准的数值。实践经验表明，释放量远小于预期。而异常释放可以立即被检测到。

主要的环境影响来源于核燃料循环前端。

应由辐射防护和健康防护机构负责对人员所受的放射性和化学影响进行评估。这类评估应根据可靠的科学数据和经过验证的受照模型进行。但是，需要持续进行研发，以降低数据的不确定性并改进模型。

在对电离辐射产生的生态系统影响进行评估时，未能获得足够的支持，需要加大这方面的研发力度。

在处置库关闭后，在监护期间将继续对其进行监视；然后其安全性将从主动安全变为被动安全。大多数放射性核素将在处置库中衰变，那些可能返回生物圈的放射性核素的衰变时间已经很长，其放射性毒性影响可以忽略不计。

目前，美国新墨西哥州的废物隔离示范设施（WIPP）已经开始运行。WIPP 位于深层盐岩地质体中，用于接收核武器研发和生产过程中产生的超铀废物。目前国际上尚无用于处置商用核电厂高放废物的深地质处置库。2015 年，芬兰颁发了世界上第一个高放废物处置库（ONKALO）建造许可证。目前，瑞典和法国的处置库项目已经进入许可审查阶段。

在社会意义方面，必须以透明的方式让公众注意到放射性废物管理对环境的现有或潜在影响。

# 附录 3　放射性废物处置设施

## A3.1　法国

法国放射性废物管理的环境应对措施经验具有指导意义。这一信息来源于三类处置库的监测：第一类目前已经关闭，第二类目前正在运行，第三类待建。

### 1. CSM

用于处置 LIL-SLW 的地表处置库 CSM（Centre de Stockage de la Manche）已投入使用长达 25 年（1969～1994 年），且临时多屏障覆盖的建造工程持续了 6 年。该处置库位于海边，建造在片麻岩基底上。该处置库共存放有 1470000 个废物包（530000 $m^3$ 的废物），占地超过 15 公顷。已对容纳废物包的土木工程结构进行了试验。自 2000 年以来，CSM 一直处于监测之下（对覆盖和单元结构的封隔测试，以及对地表和地下出水量的测试（每年进行 10000 次测量））。在完成屏障覆盖建造后，该设施将被关闭，预计在 2050/2060 年左右关闭。对 CSM 的监测将持续进行，直至实现被动安全。根据对 CSM 监测的反馈，氚较难封隔。在充分利用处置库附近水流的情况下，最大辐射影响为 0.20 μSv/a。

**2. CSA**

用于处置 LIL-SLW 的地表处置库 CSA（Centre de Stockage de l'Aube）自 1992 年投入使用，占地 95 公顷。该处置库位于苏莱内迪伊（奥布省）附近，建造在砂和黏土层上。该处置库可堆积 $10^6$ $m^3$ 的废物。处置库现场及附近的大多数暴露组人群的辐射照射剂量不超过 0.25 μSv/a。CSA 的管理工作借鉴了 CSM 的经验（见上文）。标准废物包将被放置在防风雨的处置单元中。当处置单元装满后，用混凝土装填废物包以形成一个整体。将从辐射、物理化学和生态参数方面对 CSA 进行监测（每年检查 15000 次）。对环境中一些放射性核素释放量的监测表明，辐射影响约为 $10^{-3}$ μSv/a。在处置场地附近或地下水中均未检测到氚。

**3. Cigeo**

在未来几年，法国将建成运行 Bure（默兹/上马恩省）附近的 IL-LLW 和 HLW 深层地质处置库 Cigeo。废物包将被放置于地表以下 500 m 建造的工程结构中，其位于厚约 130 m 的黏土层中。地表设施在接收这些废物包后，对其进行整备以便于地下运输。预计 Cigeo 可运营 150 年。在此期间，将对 Cigeo 进行环境监测。为此，Andra 自 2007 年开始在 Cigeo 设施附近运营一个观察范围达 900 $km^2$（参考面积 250 $km^2$）的永久环境观测站（environmental permanent observatory，EPO）（1.5 km×1.5 km 的网格）。此观测站旨在：①测量与人类活动有关的物理化学交换，了解对空气、水、土壤、动植物的所有影响；②记录和保存这些数据。EPO 还有一个附属的样品保存设施。EPO 设有 2500 个观测点，每年可收集 2500 份样本和 85000 个数据样本。

## A3.2 中国

### 西北处置场

西北处置场位于中国西北地区的甘肃省，采用近地表处置方式接收、贮存和处置低中放固体废物。西北处置场于 1995 年开工建设，1998 年建成，2011 年获批运行许可证。西北处置场规划处置容量 20 万立方米，首期工程处置容量为 6 万立方米，由 17 个处置单元构成。目前已建成 2 万立方米，6 个单元。西北处置场位于厚度约为

50 m 的黏性土和砂性土互层中，处置单元采用了加固水泥结构，废物桶之间和废物桶与处置单元壁之间用砂土回填，处置单元装满后浇筑钢筋混凝土顶板。处置库关闭时，在处置单元上铺设 2 m 厚的最终覆盖层。在处置场建造过程中，为了进一步提高其安全性，处置单元均增加了混凝土底板。在对处置场监护期满后公众无意闯入情景进行评估时，考虑了公众在处置场上居住、钻探、打井饮水等景象。分析结果表明，在这些景象下，无意闯入者受到的剂量不会超过 0.1 mSv/a，远低于国家对公众年有效剂量 1 mSv 的限值。截至 2019 年 12 月 31 日，西北处置场已经接收了约 2.2 万立方米的废物。

# 第4章　严重核事故

---

建议

● 尽可能加大严重事故机理研究的投入，对严重事故缓解提供支持。进一步开展保持安全壳完整性的措施研究以及先进技术（如ATF燃料等）的开发应用研究。

● 进一步积累核电厂严重事故管理导则的实施经验，积极推进大范围损伤缓解和多机组事故管理措施的制定，并努力提高事故应急响应能力。

---

## 本章介绍

自20世纪50年代和平利用核能以来，经多年发展，核电已经与火电、水电并称为全球电力供应的三大支柱。全球共有454台核电机组，整体安全状况良好（国际原子能机构动力堆信息系统（Power Reactor Information System，PRIS）数据）。商用核反应堆50年来的正常运行证明，其辐射影响极低，且远低于天然辐射水平（参见中法关于核能未来的联合建议报告）。然而，在核电发展的历史上，也发生了三里岛、切尔诺贝利和福岛这样的严重核事故，对核能发展和世界人民对核能的认知造成了巨大影响。因此，有必要回顾这三次核事故，深刻总结事故经验教训，梳理事故后核能行业在降低严重事故发生概率及其后果方面所采取的设计改进和措施，这对核能的未来发展非常重要。

本章首先回顾了三次严重核事故，并说明了它们的影响以及吸取的经验教训（4.1节）。4.2节讨论了通过不断改进技术和管理来避免此类事故及限制事故后果的措施。4.3节给出了本章主要结论。此外，为进一步阐明当前中国针对核电厂严重事故所作的技术与政策考虑，本章最后提供了三个资料附录，分别介绍：第三代压水堆严重事故专用预防和缓解措施（附录 4－1），中国内陆核电厂的安全性（附录 4－2）以及中国事故后应急组织管理（附录 4－3）。

## 4.1　核电厂严重事故

影响世界核工业发展的重大事故有三次：1979 年发生在美国的三里岛核电厂事故；1986 年发生在苏联的切尔诺贝利核电厂事故；2011 年发生在日本的福岛核电厂事故。三里岛核电厂事故在国际原子能机构核与辐射事故分级表（IAEA INES）中定级为 5 级，代表对环境无影响或造成了有限影响。后两起事故定级为 7 级，对环境造成了严重影响。

这些事故的成因各异，对环境的影响和所吸取的教训也不同。但是，事故的严重后果极大促进了核能安全技术的进步，提高了核安全水平。因此，即使再次发生严重事故，仍然可以明显降低人们所担心的环境影响。当前，全球在建核电机组均进行了明显的设计改进，属于第三代核电技术；并且随着运行经验的不断积累，核电机组的运行管理能力均得到有效提升；即使最坏的情况下，放射性物质向环境释放的风险也降低到了非常低的水平。与此同时，安全监管当局也发布了事故应急和补救指南，相关法规还强制要求营运单位承担实施这些指南的义务。

本节分析了这三起事故及其在技术和管理水平与法规制度方面引起的变化。每一起事故的分析重点是，如何减少核能对环境的影响。

### 4.1.1　三里岛核电厂事故[34]

#### 1. 事故原因

三里岛核电厂采用美国早期的压水堆机型，事故起因包括设备故障、操作人员对系统状态了解不充分以及随后的决策不当。导致堆芯熔化，大量裂变产物进入安全壳。幸运的是，安全壳保持完好，有效包容了事故产生的放射性物质。

#### 2. 环境影响

这次事故向环境释放的辐射非常有限。对周围公众的最大剂量低于年度天然本底剂量，仅为后者的 1/10。没有造成人员伤亡，也没有对环境造成中长期的影响。

3. 经验教训

尽管三里岛核电厂事故仅对环境和公众造成了微小的放射性后果，没有导致人员伤亡，但直接经济损失巨大。此次事故为美国核工业以及监管当局敲响了警钟，并对世界核工业的发展产生了深远影响。三里岛核电厂事故也使当时快速发展核工业的欧洲和其他国家重新思考核电安全问题。

虽然整体安全防御效果出色，但三里岛核电厂事故仍暴露出设计、管理和安全研究方面的缺陷和不足。事故表明，之前未予重视的小故障居然会酿成如此严重的后果。事故反映出管理方面（例如操作员培训、应急程序、组织和协调）的重要性并不亚于技术方面（例如设备设计、建造、鉴定和安全分析），对核电厂安全运行同样具有不可忽视的影响。

自从三里岛核电厂事故发生以来，核工业界在人机界面、监测控制、人员培训等方面进行了许多改进并取得了重大进展。在事故安全分析领域发生了重大变革，核工业企业和监管机构协同行动，将重点从以设计基准事故（design basis accident，DBA）分析为目标的反应堆安全研究转变为反应堆的严重事故研究，并投入巨资开展了大量的严重事故研究项目。作为一个分水岭，安全分析研究的重点由大破口失水事故（large break loss of coolant accident，LBLOCA）转向小破口失水事故（SBLOCA，small break loss of coolant accident）和瞬态研究。在 Rasmussen 教授领导的研究组于 1975 年发表的 WASH-1400 报告中，采用的概率风险评价技术（probabilistic risk assessment，PRA）体现出传统确定论分析技术所无法比拟的科学预见性。三里岛核电厂事故作为一个标志性事件，使 PRA 技术得到了重生，并在随后的发展应用中展现了强大的生命力。

### 4.1.2 切尔诺贝利核电厂事故

1. 事故原因[35]

切尔诺贝利核电厂事故发生的原因之一与 RBMK 反应堆的固有特性有关。该反应堆是石墨慢化水冷堆，具有正的空泡系数，存在瞬发超临界的潜在风险，并且没有包容放射性物质的安全壳。操作员对系统状态了解不充分，违规操作导致瞬发超临界事故，反应堆功率急剧上升，引发爆炸，造成大量放射性物质泄漏到大气中。切尔诺贝利核电厂事故发生的原因主要是设计缺陷、运行管理混乱和安全文化缺失。

发生事故的切尔诺贝利反应堆是由苏联设计和建造的17个RBMK反应堆（仅在苏联部署）之一，事故后已不再建设。

### 2. 环境影响

切尔诺贝利核电厂4号机组的主要放射性释放持续了10天，欧洲的大部分地区均在一定程度上受到了切尔诺贝利释放的影响。大部分释放物是短半衰期的放射性核素，长半衰期的核素释放量较小[36]。

134名应急工作人员罹患急性放射病，其中28人死于辐射。在受到中等剂量照射的恢复清理工作人员中，发现了白血病和白内障危险增加的证据。在儿童和青年期受到照射的人群中，甲状腺癌的发生显著增加，其归因于在事故早期饮用了放射性碘污染的牛奶[37—39]。

事故后6个月内完成了石棺的建造，其目的是对受损的反应堆进行环境控制，降低现场的辐射水平，防止放射性进一步外泄[36]。事故后，该核电厂持续监测放射性水平，定期接受国际审查以评估形势的发展。与此同时，周围的自然生命持续发展。目前，有关部门正在研究评估事故是否造成植物和野生动物的基因影响。

### 3. 经验教训

切尔诺贝利核电厂事故及事故后处理工作给苏联、乌克兰、白俄罗斯和俄罗斯带来了巨大的经济损失与财政负担，对当地人口、当地活动和全世界的核工业造成了深远影响。这起事故引发了公众对核安全的关注，迟滞了核能设施的规划，事故影响的深度和广度远超三里岛核电厂事故。

切尔诺贝利核电厂事故给核工业界提供了相当多的经验教训：

（1）事故发生后，业内几乎彻底摒弃了带有正反馈效应的堆芯核设计方案，石墨反应堆的发展至此告终，反应堆的固有安全属性得到了增强。

（2）反应堆保护系统得到改进，增加对主控室人员的操作限制，有效降低核电厂发生人因失误的可能性。

（3）承压安全壳作为最后一道安全屏障成为全行业的共识，因其能够进一步降低大量放射性物质释放可能性，保护公众健康和环境安全。

(4) 安全文化应运而生，受到全世界核行业的高度重视；核安全意识广泛传播，贯穿核电厂的设计、制造、建造、运行、监管等诸多领域，为预防核事故发生发挥着重要作用。

(5) 冷战背景下核技术领域的意识形态隔绝被打破。国际原子能机构制定并实施了《核安全公约》，要求监管机构开展国际同行评估。世界核营运者协会（World Association of Nuclear Operators，WANO）等国际组织成立，鼓励营运单位加强操作运行安全，培养“核安全无国界”理念。

尽管切尔诺贝利核电厂事故后果严重，影响深远，但从事故的后续研究中也可以看出，只要严格遵守安全准则，增强安全意识，不断合理优化设计，使新的核电厂具有高度的固有安全性，核电安全依然可以得到保障。

### 4.1.3 福岛第一核电厂事故[40, 41]

#### 1. 事故原因

日本福岛处于欧亚板块与太平洋板块“俯冲带”附近，历史上大地震频发；福岛第一核电厂采用美国最早的商用核电厂技术——沸水堆（boiling water reactor，BWR)，其设计与建造完成于美国三里岛核电厂事故之前（当时对严重事故以及复杂事故序列还没有清晰的认识）。预防和缓解严重事故的安全措施在设计上“先天不足”。

福岛第一核电厂事故的主要原因是自然因素，超强地震引发超强海啸远超设计标准。9 级地震及海啸对周边地区交通、电力等基础设施造成的严重破坏导致场外电源直到地震后 9 天才恢复，时间之长远超设计考虑范围。在福岛第一核电厂六个反应堆中，有四个在事故中损坏，其均位于海岸附近，设计初衷是尽量减少冷却回路的长度。这一设计不仅增加了海啸袭击风险，还使作为场外电源的柴油发电机组失去作用。柴油发电机房的气密性不足，进气口被水淹没，最终导致所有交流电丧失，无法进行堆芯冷却而发生灾难；并且还缺乏消除氢气的装置，导致爆炸，将放射性物质送入大气层。此外，无法对乏燃料池进行冷却也导致向大气排放了更多的放射性物质。

#### 2. 环境影响

福岛第一核电厂事故导致放射性气体（少于 500 PBq（P 代表 $10^{15}$）的放射性碘，

少于 20 PBq 的放射性铯）和其他放射性物质释放入环境。虽然在事故早期阶段，剂量率超过了一些参考值，但预计不会对动植物种群和生态系统产生任何影响。鉴于估计的短期剂量一般远远低于能够产生高度有害后果的严重水平，并且剂量率在事故后的下降速度较快，因此预计不会产生长期效应。

在场址 20 km 半径内和其他指定区域的居民被撤离，在 20 km 至 30 km 半径内的居民则被指示进行掩避，后被建议自愿撤离。很多人因撤离失去了农场和企业。日本政府已进行了大规模的修复工程来清理这一地区，以便人们能逐步重新回归原有土地。

### 3. 经验教训

福岛第一核电厂设计中的严重事故预防和缓解措施不足也是事故的重要原因之一。如前所述，福岛反应堆于三里岛事故前完成了设计和建造，当时对严重事故和复杂事故序列没有明确的认识。在这些早期的反应堆系统中，存在事故预防和缓解系统的设计缺陷。

事故的经验教训为新建核电厂的设计改进和加强在运核电厂的运营管理提供了参考。对于福岛核电厂事故，业界主要提出了以下反馈：

（1）在核电厂的设计和运行中需要对超设计基准的外部事件加强考虑，对自然危害的评估需要足够保守。对自然危害的评估需要考虑其先后发生或同时并发的可能性及其对核电厂的综合影响。

（2）强调应急泵的绝对密封性。

（3）重点关注乏燃料水池的结构完整性和长期冷却问题。

（4）定期评估核电厂的安全性，将最近技术、必要的纠正行动或需要立即实施的补偿措施结合在内。

（5）必须在设计基准事故期间使仪器仪表和控制系统保持可用，以便监测核电厂基本安全参数并为核电厂运行提供便利。在核电厂余热排出方面，需要提供在设计基准事故工况和超设计基准事故（beyond design basis accident，BDBA）工况下能够稳固可靠运行的冷却系统。

（6）培训、演习和演练应包括假想严重事故工况，以确保操作人员做好充分的行动和决策准备。

事故发生后，国际原子能机构不仅与日本核能与工业安全局（Nuclear and Industrial Safety Agency，NISA）建立了直接联系，而且持续为成员国、有关国际组织和公众提供更新和发布最新信息。

福岛核电厂事故对日本造成的伤害程度和对世界环境的影响至今尚难定论，但如同切尔诺贝利核电厂事故一样，它形成的冲击波，让整个世界绷紧了神经，对发展核电可能造成的环境灾难日趋忧虑。

事故发生后，世界有关国家对本国核电安全性重新进行了评估。经过检查和评估，许多国家确认了继续发展核电的立场，并采取措施改善现有核电厂的安全状况，提高应急反应能力。福岛核电厂事故虽然减缓了世界核能发展的速度，但同时也将进一步促进核电安全和管理水平的提升与发展。

## 4.2 为降低事故环境影响采取的改进措施

综上所述，三里岛核电厂事故后，世界各国在压水堆核电厂的设备可靠性、操作员培训、人机界面等方面做出了一系列重大改进。切尔诺贝利核电厂事故后，各国为提高核电厂的安全性进行大量研究，并在此基础上开发出先进的核电技术。福岛第一核电厂事故后，各国组织了全方位的核电安全检查并进行了改造升级。

### 4.2.1 反应堆技术提升

鉴于上述事故的教训，业界对在运和在建的第二代压水堆核电厂实施了许多重要技术改进。应“提高核电厂安全性、可用性、可靠性和经济性”的全面要求，核工业界提出了第三代压水堆核电厂的概念，从而实际消除严重事故后大量放射性释放。

“实际消除”概念最早由欧洲提出，后被国际原子能机构（IAEA）采纳。中国核工业界关于“实际消除”定义的认知已基本取得一致，即：如果某些工况物理上不可能发生，或以高置信度认为某些工况极不可能发生，则可以认为这些工况发生的可能性已被实际消除。中国新建核电机组力争实现从设计上实际消除大量放射性物质释放的可能性。中国政府在核安全和放射性污染防治相关规划文件中明确了这一目标。

第三代压水堆核电厂贯彻纵深防御理念，设置多道实体屏障，并且采用多重性、多样性与实体隔离的设计原则来提高事故发生时的响应能力和缓解能力，实际消除严重事故后的大量放射性释放。预防和缓解严重事故的具体目标包括：

（1）防止堆芯熔化；

（2）保持反应堆压力容器（RPV）的完整性；

（3）保持安全壳的完整性；

（4）防止乏燃料的放射性释放。

第三代压水堆均设置了先进的大型安全壳，能够承受地震、龙卷风等外部自然灾害和火灾、爆炸等人为事故的破坏与袭击，以及大型商用飞机恶意撞击或恐怖袭击；耐受严重事故情况下所产生的内部高温高压、高辐射等环境条件，并保持其完整性，避免放射性物质向环境释放。附录4－1给出了中国核工业界为防止大量放射性释放而采取的对策和特殊改进措施。

法国和中国的核工业界均已发展出了各自的第三代压水堆技术，分别为EPR和“华龙一号”（HPR1000），其中EPR首堆工程已并网发电，HPR1000首堆工程进展顺利。

另外，中国核工业界对公众特别关注的“内陆核电”安全性问题做了针对性的技术改进，如正常运行时放射性液态流出物的“近零排放”、严重事故工况放射性废液的处理等，安全水平满足了当前国际最高安全标准要求，详见附录4－2。

为进一步提高核电厂的安全性能，特别是压水堆核电厂安全性能，核工业界积极开发新型技术，例如开发新一代事故耐受燃料（ATF），以最大限度地减少事故期间可能产生的氢气，从源头消除事故期间发生氢气爆炸的可能性。严重事故后堆内熔融物的滞留技术也是业界的关注重点之一，多家研究单位在强化传热、提高临界热流密度（如纳米流体的应用）等方面开展了相关工作。

### 4.2.2　核电厂在福岛第一核电厂事故后的行动

福岛第一核电厂事故后，中国大陆和西方国家核电厂运营企业积极开展核安全自查。同时，根据监管机构发布的技术要求实施了多个方面的技术改进，包括核电厂防洪能力改进、应急补水及相关设备改进、移动电源及设置改进、乏燃料水池监测改进、氢气监测与控制系统改进、应急控制中心可居留性及功能改进、辐射环境监测及应急改进、外部自然灾害应对改进等。

以核电厂防洪能力改进为例，核电厂会进行水淹再评估，计算出厂址发生设计基准洪水同时遭遇千年一遇降雨的厂区最大积水深度。将确定后的厂区最大积水深度作为相关构筑物、厂房防水封堵的设计依据。结合核电厂可能遭受水淹情况的评估，分别对核安全相关厂房的廊道、门窗、管沟和贯穿件等防水封堵措施进行核查，并加强改进其中薄弱环节。通过增加孔洞防水封堵功能、封堵廊道与核岛的接口，确保核电厂在满足外部水淹设计基准防护要求的基础上，进一步加强抵御超设计基准外部水淹的能力。

以上技术改进的实施增强了在役核电厂应对包括多重失效在内的超设计基准事故能力，防止类似于福岛第一核电厂事故的发生。

法国核安全局根据法国政府的要求，针对全国核设施发布了进一步安全审查技术要求，重点涉及以下项目：洪水、地震、断电、降温、事故管理、技术评估和现场核查。

### 4.2.3 严重事故管理

除了提升核电厂的运行管理之外，国家和国际两级还建立了两套技术导则，用于严重事故的缓解和应急。

**1. 严重事故管理导则（severe accident management guidelines，SAMG）**

核电供应商针对不同类型的电厂设计制定了一系列 SAMG 导则或框架。第一份框架文件 WOG SAMG（WOG，Westinghouse Owner Group，为西屋所有者集团）由美国西屋公司于 1994 年发布，其制定依据包括严重事故研究结果，以及美国电力科学研究院（Electric Power Research Institute，EPRI）制定的技术基础报告（technical basis report，TBR）。这份文件总结了美国典型压水堆严重事故管理研究的进展。因具有技术先进、逻辑性强、功能体系完善等优点，WOG SAMG 在国际上得到广泛认可及应用。

2019 年，国际原子能机构发布了核安全导则 SSG-54《核动力厂事故管理》。该导则对预防和缓解严重事故后果的管理程序制定提出了明确要求。

自第一版以来，WOG SAMG 导则的框架体系和内容方面不断改进，并于 2016 年在 WOG SAMG 的应用基础上，完成了全新体系的 PWROG SAMG。它结合了 20

多年来在严重事故管理领域的最新研究成果和工程经验，从而使其具有了更加广泛的适用性和便捷性。

法国 EDF 制定了主要针对法国第二代压水堆核电厂的严重事故管理指南（Guide d'Intervention en Accident Grave，GIAG）。法玛通还为 EPR 的严重事故管理制定了 OSSA 指南。OSSA 指南涵盖了电厂的所有阶段，包括额定功率运行、低功率运行、停堆和乏燃料水池存储。

从 21 世纪初开始，中国国内开始制定和实施 SAMG。截止目前，中国所有在运核电厂均完成了严重事故管理导则的制定，并且还要求所有在建核电厂制定严重事故管理导则。福岛核电厂事故后，绝大部分核电厂的严重事故管理导则已经大大扩展，涵盖额定功率运行、低功率运行、停堆以及乏燃料贮存设施，但由于制定单位不同及理念各异，导致具体的导则框架结构存在一定的差异。

2013 年起，受中国国家核安全局委托，中国核能行业协会组织业内专家对多个核电厂的 SAMG 开展了同行评估，包括田湾核电厂 1、2 号机组、岭澳核电厂、方家山核电厂和秦山三期等，以检查和评估 SAMG 在各核电厂的制定和实施情况。在国家核安全局的要求下，核电厂也均定期开展 SAMG 培训，并将 SAMG 的实施纳入核电厂的演习计划中，作为核电厂综合演习的一部分，或实施严重事故专项演习。

SAMG 的制定和优化、同行专家的评估活动以及核电厂的专项演习等活动有效提高了中国核电厂应对严重事故的管理和应急水平，大大提高了核电厂的运行安全性。但由于国内核电厂 SAMG 的制定和实施时间较短，经验还相对欠缺，在相互配合、多机组事故管理、管理导则验证等方面还需要进一步研究。

**2. 大范围损伤缓解导则（extensive damage mitigation guidelines，EDMG）**

对于核电厂，需要考虑恐怖袭击造成的火灾和爆炸等情况产生的极端破坏。核电厂可能会因这种情况发生大范围损伤，导致常规的事故管理程序失效。2001 年 9 月 11 日，美国发生纽约世贸中心双子塔恐怖袭击事件。之后，美国核管会要求核电厂针对重大火灾和爆炸导致的电厂大范围损伤制定事故管理策略和导则，以维持和恢复堆芯冷却、安全壳的完整性并保证乏燃料池冷却。

美国核电厂已根据美国联邦法规 10CFR50.54 的要求制定了大范围损伤缓解导则（EDMG），以应对重大火灾或爆炸造成的大范围损伤。EDMG 的边界条件为主控室无法进入，在主控室与远程停堆站丧失对核电厂状态的控制能力导致 EOP 与 SAMG

无法使用时，需要采用 EDMG 作为响应导则。因爆炸、火灾、飞机坠毁等造成核电厂大面积损伤的情况下，核安全许可证持有者应当就保证或恢复堆芯冷却、安全壳和乏燃料池冷却能力等情况制定并实施相关策略。

欧洲则开展了对核电厂的“压力测试”，主要评估极端外部事件对核设施的影响，分析恶意或恐怖行动导致的安保威胁与反应堆事故。在开展评估后，法国电力公司（EDF）提出，应制定“核能快速行动救援力量”（La Force d'Action Rapide Nucléaire，FARN）计划，由国家级专业人员配备先进设备组成应急行动队，作为应急响应的补充，为事故现场快速提供应急与物资的支持，并具备多机组同时干预的能力。FARN 计划需要明确启动准则、可能的任务、应急专业人员及应急资源的配置、人员培训要求，以及核电厂的相应管理程序。南非、西班牙等国的核电厂已经制定了 EDMG；韩国也正在研究制定 EDMG，以应对极端外部灾害造成的大范围损伤。

近年来，中国国内各研究院所也在积极开展大范围损伤缓解导则的研究与制定工作。目前，红沿河、方家山、三门核电厂 1&2 机组、福清核电厂 5&6 号机组等已经完成 EDMG 的制定，其他核电厂也正在积极推进。国内制定的 EDMG 主要考虑主控室与远程停堆站在丧失对核电厂状态的控制能力，导致 EOP 与 SAMG 无法使用的情况下，如何进行事故管理和应急。在国家核安全局的组织下，国内也曾多次组织、召开研讨会，讨论 EDMG 的制定和实施，以及对核电厂应急计划的影响。总的来说，国内在 EDMG 制定和实施方面已经迈出了有益的探索步伐，但制定和实施经验非常欠缺，亟需进一步研究 EDMG 在核电厂的实施以及与现有应急体系的融合。

### 4.2.4 对于未来类似严重事故的见解

在已经发生的三次严重核事故中，真正造成严重后果的是切尔诺贝利核电厂事故和福岛第一核电厂事故。

切尔诺贝利事故的原因是，反应堆设计缺陷和操作人员多次违反安全操作规程，致使反应堆失去控制和保护，引入超瞬发临界反应性，使得功率剧增引起反应堆爆炸。事故后，业内几乎彻底摒弃了带有正反馈效应的堆芯核设计方案，反应堆的固有安全属性得到了增强。

由大地震引发海啸而导致的福岛第一核电厂事故，是历史上第一次由外部灾害引发的核电厂事故，也是人类史上继切尔诺贝利核电厂事故后的第二个被评为国际核与辐射事件分级（INES）7 级的核事故。

福岛第一核电厂事故后，中国进行了沿海核电厂地震海啸安全性分析。

中国属于欧亚大陆板块，大地构造上属于板块内部地区，主要的破坏性地震活动为大陆板块内部及地壳内部的浅源地震，与板块俯冲带产生的地震相比能量要小很多，地震产生的形变位移远远达不到产生海啸的条件。此外，中国海域有宽缓的大陆架，水深条件不利于海啸能量的积累。因此，无论从地震的震源条件还是海域的水深条件，都反映出中国沿海与日本完全不同，因而中国沿海不具备发生类似日本那样的大规模地震海啸的条件。法国也是如此，千百年来，这种极端的自然灾害从未被观测到。

另外，使福岛第一核电厂事故如此严重的几个特征（应急泵被淹没、缺少消氢措施等）都通过 4.2.2 节所述的技术改进予以消除。

## 4.3　结论

今后可能发生严重事故的环境风险已大大降低。在运和在建核电厂配备有完善的严重事故缓解措施，能够实现放射性源项排放可控，并限制事故的影响程度和范围。事故仅会造成有限区域的影响，即无需永久迁居、核电厂周边地区无需紧急撤离、只需为有限的人员提供庇护所、无需长期限制食品消费。

完善的严重事故预防与缓解措施，使得第三代反应堆安全等级更高。第三代压水堆均设置了先进大型安全壳，能够承受地震、龙卷风等外部自然灾害和火灾、爆炸等人为事故的破坏与袭击，以及大型商用飞机恶意撞击；耐受严重事故情况下所产生的内部高温高压、高辐射等环境条件，并保持其完整性，避免放射性物质向环境释放。

无论从堆型、自然灾害发生条件还是安全保障方面来看，类似切尔诺贝利和福岛第一核电厂事故这类造成大量放射性物质释放的事故都不可能在中国、法国发生。

但事故是不可精确预测的。一般来说，未来发生的任何事故都将不同于以往的事故。控制风险的理性的方法是考虑意外情况下可能发生的事故，所以要增强缓解功能，尽可能减少潜在的场外后果，不影响其他的人类活动。

# 附录 4-1　第三代压水堆严重事故专用预防和缓解措施

针对安全壳内的各类严重事故现象，第三代压水堆核电厂设置了一系列严重事故缓解措施，以保持安全壳完整性，防止放射性物质大量释放。目前，威胁压水堆核电厂安全壳完整性的主要严重事故现象及其应对措施总结如下：

（1）为了避免高压熔堆引起的安全壳直接加热（direct containment heating，DCH），一般在稳压器上设置专用快速卸压阀。

（2）采用消氢系统控制安全壳内的可燃气体。

（3）为了避免安全壳超压风险，增加安全壳热量排出系统的可靠性和冗余度，包括增加安全壳喷淋设施数量和保证可靠水源，采用非能动的热量导出系统等。

（4）为了避免压力容器外蒸汽爆炸现象，一般采用干堆坑设计，或者反应堆压力容器外部冷却的措施，防止压力容器熔穿。前者排除蒸汽爆炸所需水源，后者阻止熔融物释放，二者均可从根本上消除蒸汽爆炸发生的可能。

（5）为了避免发生堆芯熔融物-混凝土相互作用（molten core concrete interaction，MCCI）和安全壳底板熔穿，可以采用熔融物堆内滞留措施或堆芯熔融物捕集器。前者避免压力容器熔穿，后者在堆芯熔融物熔穿压力容器后把其收集起来并冷却。

（6）针对安全壳的旁路失效，电厂主要采取增加隔离可靠性、提高低压系统设计压力等措施。为了减小界面冷却剂丧失事故（loss of coolant accident，LOCA）引起的严重事故的释放量，要求与反应堆冷却剂系统相连的正常工况使用的系统，必须位于有包容功能的厂房内；与冷却剂系统相连的事故工况下使用的系统，应尽量设置于安全壳内。针对蒸汽发生器传热管破裂（seam generator tube rupture，SGTR）旁路释放，采取必要的措施，防止 SGTR 下蒸汽发生器发生满溢。

特别是，在部分严重事故缓解措施失效、安全壳内压力持续升高的情况下，过滤排放系统能够实现安全壳卸压和可控排放。过滤排放的理念是通过主动卸压使安全壳内的压力不超过其承载限制，从而确保安全壳的完整性。同时，通过安装在卸压管线上的过滤装置对排放气体的放射性物质进行过滤，针对气溶胶的过滤效率能够达到 99.9%，仅有惰性气体、少量挥发性物质能够排放到环境中。

# 附录4-2 中国内陆核电厂的安全性

对于沿海（滨海）核电与内陆（滨河、滨湖）核电的问题，世界各国核电厂除各厂址对工程适应性的普遍要求外，都执行统一的核安全审评标准，遵守统一的建造、运营规范，并不存在所谓的内陆核电争议问题。目前，世界在运核电机组有超过半数分布在内陆地区，其中：美国在运核电机组99台，74%位于内陆；法国在运核电机组58台，70%位于内陆。中国大陆目前则存在着内陆核电的暂时性管理问题。

中国核与辐射安全法规标准依据IAEA的安全标准制定，并持续改进、不断提高、保持与国际接轨，依据中国现行核安全法规进行内陆核电建设，满足当前国际核电建设最高安全标准。

中国大陆核电厂厂址设计基准地震的确定采用了国际上严格的标准，同时考虑极端情况下的地震影响与千年一遇的抗震设防标准。目前论证充分的内陆厂址地震加速度峰值均小于0.15$g$，而第三代压水堆核电厂抗震设防的设计标准为0.3$g$，表明上述厂址的第三代压水堆核电厂的地震安全得到充分保障。另外，内陆核电采用“干厂址”选址理念，确保核电厂免受洪水淹没的影响。在冷却水方面，内陆厂址循环冷却水采用冷却塔闭式循环，耗水量较少，且不会由于温排水而对取排水水域带来“热污染”。

内陆核电厂厂址选择必须严格按照现行核安全法规，满足有关气态、液态流出物和人口分布的相关要求。而且，内陆核电厂液态流出物排放标准比沿海更为严格，其浓度限值较沿海低一个数量级。核电厂采用从源头控制放射性废物产生的设计，应用最佳可行技术进行放射性流出物处理，通过严格的放射性废物监测、优化排放管理、加强环境监测等综合措施，实现放射性液态流出物的“近零”排放。

如4.2.1节所述，中国内陆核电厂址采用的第三代压水堆核电技术设有完善的严重事故预防和缓解措施，能够有效预防严重事故的发生和缓解严重事故的后果。针对严重事故后的放射性废水处理问题，中国核工业界开展了研究。即使发生极端严重事故，第三代核电厂设计能够产生的最大的放射性污水总量约为7000～10000 $m^3$。为防止这些放射性污水污染周围环境水体，设计中采取一系列措施，包括：反应堆厂房、核辅助厂房等安全厂房的放射性污水贮存；配备多台大容量的废液贮罐和临时废液暂存池，作为安全厂房废液贮存能力的补充或后备；设置防止泄漏的阻水剂、放射性污染物抑制剂，沸石过滤装置等，以备在紧急情况下使用，实现放射性废水的封堵及与地表水体间的隔离；厂址区域内预留足够空间，保证能够将产生的废液及时通过移动式应急废液处理装置处置。通过上述措施，即使在极端情况下，也能实现放射性废液的“贮存、处理、封堵、隔离”，保障核电厂即使发生极端严重事故，放射性释放对环境的影响也是可控的，保障环境安全。

综上所述，目前中国核电厂采用的核与辐射安全标准与世界最高安全标准接轨，采用技术成熟、符合第三代安全标准的核电技术，其安全水平满足当前国际最高安全标准要求。只要严格遵守核安全法规标准，并采取合理有效的工程措施，中国内陆核电厂厂址安全就能够得到有效保障，正常运行工况下对公众和环境影响在可以接受的天然本底涨落范围内，严重事故工况下核电厂环境风险可控（无需永久迁居、核电厂周边地区无需紧急撤离、只需为有限的人员提供庇护所、无需长期限制食品消费）。

## 附录 4 - 3　中国严重事故后应急组织管理

根据中国《国家核应急预案（2013）》，严重事故发生后，各级核应急组织根据事故的性质和严重程度，实施以下全部或部分响应行动：

（1）事故缓解和控制；

（2）辐射监测和后果评估；

（3）人员放射性照射防护；

（4）去污洗消和医疗救治；

（5）出入通道和口岸控制；

（6）市场监管和调控；

（7）维护社会治安；

（8）信息报告和发布；

（9）国际通报和援助。

开展事故现场和周边环境（包括空中、陆地、水体、大气、农作物、食品和饮水等）放射性监测，以及应急工作人员和公众受照剂量的监测等。实时开展气象、水文、地质、地震等观（监）测预报；开展事故工况诊断和释放源项分析，研判事故发展趋势，评估辐射后果，判定受影响区域范围，为应急决策提供技术支持。

### A4－3.1　中国的三级核应急体系

国家核事故应急协调委员会由核工程与核技术、核安全、辐射监测、辐射防护、环境保护、交通运输、医学、气象学、海洋学、应急管理、公共宣传等方面专家组成，为国家核应急工作重大决策和重要规划以及核事故应对工作提供咨询和建议。

省级人民政府根据有关规定和工作需要成立省（自治区、直辖市）核应急委员会，由有关职能部门、相关市县、核设施营运单位的负责人员组成，负责本行政区域核事故应急准备与应急处置工作，统一指挥本行政区域核事故场外应急响应行动。省核应急委员会设立专家组，提供决策咨询；设立省核事故应急办公室，承担省核应急委员会的日常工作。此外，还设立省核应急前沿指挥部，为事故后核应急指挥提供决策支持。

核设施营运单位的核应急指挥部负责组织场内核应急准备与应急处置工作，统一指挥本单位的核应急响应行动，配合和协助做好场外核应急准备与响应工作，及时提出进入场外应急状态和采取场外应急防护措施的建议。

### A4－3.2　核应急监测体系

国家核事故应急指挥部或国家核应急协调委员会视情组织国家核应急力量指导开展辐射监测，组织协调国家和地方辐射监测力量对已经或可能受核辐射影响区域的环境（包括空中、陆地、水体、大气、农作物、食品和饮水等）进行放射性监测。

省级人民政府和核电厂的核事故应急机构应当做好事故后环境放射性监测工作，为采取核事故应急对策和应急防护措施提供依据。

省环保厅下设省环境监测组，包括陆地、海上、空中、食品与饮水等监测小组。省辐射环境监测管理站配有通信、数据采集、处理与传输等设备，负责协助省环境监测组组织协调核事故场外应急监测，汇集整理所有监测数据，分析事故对环境及公众可能带来的辐射影响，为省核应急评估中心提供监测数据，为省核应急指挥部提供决策依据。

核设施运营单位应急响应组负责协调和实施应急辐射监测和环境取样，确保事故后可以迅速启动应急辐射监测。应急响应组包括监测人员、取样人员、监测/取样指导与协调人员，以及对监测和取样人员提供的数据、样品和其他信息进行分析的人员。场区内每天至少有一个受过训练的监测组能随时启动，进行应急辐射监测，一个应急辐射监测组可以单独和同时承担监测和取样职责。

目前，国家已成立中国核应急救援队，由 6 支救援分队组成，规模约 320 人。核应急救援队是在国家核应急体制框架下依托军队及核工业现有核应急力量组建成立的；承担复杂条件下核电厂严重事故的突击抢险和紧急处置任务，有效封控核事故源头、及时搜救受困人员、全力遏制事态蔓延、最大限度减轻危害后果，支援核设施单位的现场处置行动。

实现多级应急指挥系统协调指挥，统一决策，严重事故后多部门统筹协调，应急资源快速调配。事故后数据信号及信息渠道通畅，决策手段多样化，避免单一数据来源切断后无法开展有效的决策支持。

# 第 5 章　核安全与环境

---

**建议**

由于环境保护的主要目标是消除大量放射性释放的可能性，因此建议核设施的所有者：

- 测试现有核设施抵抗超出设计基准考虑范围的外部事件的能力；
- 升级现有核设施，以达到与新建设施相同的合理可行的安全目标；
- 对所有设施实施风险指引的纵深防御，包括针对“超出设计基准工况”的应对措施；
- 对安全管理体系开展内部和独立评审，不完全依赖于安全监管部门的评审。

由于环境保护对人们来说是一个重大的敏感问题，建议核安全监管机构：

- 开展透明化的沟通，对核安全进行透明化的监督；
- 与当地政府和公众进行长期对话。

---

由于核工业的数字化进程在迅速发展，应当特别注意对设计、建造和运行阶段所使用的软件和数据库提供保护。核设施运营单位须确定一名首席安全官（chief security officer，CSO），并专门成立一个制定和实施数字化安保政策的部门，由 CSO 作为负责人。

## 本章介绍

核安全的基本目标是保护人类和环境免受电离辐射的有害影响[42]。这一目标主要是通过在核电厂和燃料循环设施的正常运行期间，控制人员的辐射照射和放射性物质向环境的释放来实现。放射性物质的释放量越小，核设施对环境的影响就越小。第2章中提到，核设施的放射性释放处于持续监测和控制下，对环境辐射水平的影响远远低于天然辐射。此外，核设施的放射性释放保持在合理可行尽量低的水平，而且记录显示，释放量随时间在不断减少，已降至核安全与环境保护机构批准允许值的很小份额。例如，在法国，液态流出物导致的平均辐射剂量低于 $10^{-6}$ Sv/a，是批准水平（$10^{-3}$ Sv/a）的 1/1000[43]，是天然辐射照射水平的 1/30。如第 3 章讨论的，废物经地质处置后长期放射性释放导致的辐射照射水平预计比天然辐射照射水平低得多。

因此，5.1 节讨论了核安全的另外两个目标：限制核事故发生的可能性和减轻此类事故发生的后果。后面三节重点探讨当今日益重要的几个特定问题：

- 选址的安全考虑；
- 政府的安全责任和作用；
- 核安全与公众接受度。

附录 5－1 介绍了核安全监管体系。

## 5.1 核电厂的安全性及其环境影响

在核能商用初期，设立的安全目标是阻止事故的发生并限制事故后果。在过去，核电厂的安全分析是基于“设计基准事故”（DBA）做出的，其旨在向监管机构证明此类事故及发生概率较高的任何事故向环境的释放相当有限。为实现这一目标，业界采取了双重方法，即：①所有用于防止发生 DBA 严重程度以下或与 DBA 同等严重的核事故的保护系统都应提供足够的冗余和多样性；②设立多重屏障以便大幅限制放射性的环境释放。该方法特别注意保护系统和屏障的独立性，被称为确定论的纵深防御，应系统实施（见 5.1.2 节风险指引的纵深防御）；在早期设计中，并未考虑堆芯熔化事故。

随着时间推移，在 WASH－1400 分析报告之后，可通过概率安全分析对确定论方法进行补充。

### 5.1.1　严重事故及其后果

第一代和第二代反应堆的设计和建造过程中运用了简化参考场景，包括根据DBA来设计安全系统及安全壳，通常使用冷却剂丧失事故（LOCA），其仅考虑了在压水堆和沸水堆的主回路中发生"双端剪切断裂"事故。然而，早在1975年人们就通过概率安全评估方法获知，DBA未能涵盖需要考虑的所有核安全事故情形[44,45]，不幸的是，发生的堆芯熔化事故（如三里岛核电厂事故（1979年）、切尔诺贝利核电厂事故（1986年）、福岛第一核电厂事故（2011年））恰恰证实了这一点。因此，需要从这些事故中吸取教训，对现有核电厂的安全规范进行调整，并修订安全目标。

在三里岛核电厂事故后，核电厂设立了定量安全目标，例如两个"千分之一"规则①。

NUREG－1150中的概率安全评估结论为：

每堆年个体早期死亡的平均概率为：

- 美国核管理委员会（Nuclear Regulatory Commission，NRC）安全目标：$5\times10^{-7}$
- 典型PWR：$2\times10^{-8}$

每堆年个体辐照致癌的平均死亡概率为：

- NRC安全目标：$2\times10^{-6}$
- 典型PWR：$2\times10^{-9}$

虽然上述结果令人满意，但是切尔诺贝利核电厂事故和福岛第一核电厂事故证明，核安全不仅应考虑核事故的致死后果，还应考虑可能要求人员撤离和重新安置（甚至包括距离核电厂较远（最远可达30 km）的人员）的环境后果。

一些国家的法规中增加了涵盖严重事故考量的最新安全目标。西欧核监管机构协会（Western European Nuclear Regulators' Association，WENRA）总结如下[46]：

- 必须实际消除会导致早期或大量放射性释放的堆芯熔化事故；
- 对于无法实际消除的堆芯熔化事故，需采取设计措施使得只需在一定区域和时

① a. 对紧邻核电厂的正常个体成员来说，因反应堆事故所导致立即死亡的风险不应该超过社会成员所面对的其他事故所导致的立即死亡风险总和的千分之一；b. 对核电厂邻近区域的公众来说，因核电厂运行所导致的癌症死亡风险不应该超过其他原因所导致癌症死亡风险总和的千分之一。

间内对公众采取有限的防护措施（无需永久迁移、核电厂邻近区域无需紧急疏散、无需长期隐蔽、无需对食品消费实施长期管控），以及有充足的时间采取防护措施。

严重事故（即堆芯熔化）可由外部灾害和/或专设安全系统故障及恐怖袭击（假设）所引发。

最重要的是，运营单位应定期检查设施抵御外部灾害（洪水、地震、飓风等）的恢复能力，确保能够抵抗比设计基准环境条件更为严重的外部灾害，确保不再发生福岛第一核电厂出现的"陡边效应"（又称"悬崖效应"，cliff edge effect）。福岛第一核电厂事故发生后，欧盟为欧洲所有核电厂进行了"压力测试"，并向公众公布测试结果。压力测试有两大益处：①有助于识别并解决一些设施的设计缺陷；②公开进行测试，向所有相关方和公众证明安全问题已得到彻底审查。目前为止，IAEA 尚未对所有核电厂的类似压力测试提供指导；对此问题发布安全导则将有助于促进测试的定期和系统实施。

针对专设安全系统，WENRA 建议对传统的 DBA 方法进行补充，要充分考虑除 DBA 以外的其他因素或扩展工况。对于这项工作，需要建立全面的风险指引的纵深防御体系方可完成。

建议对在运核电厂实施类似的安全原则，改进其安全水平，向新建核电厂设定目标看齐（实际消除大量或早期放射性释放）。为此，WENRA 提供了参考准则[47]。目前，作为大型改造项目（Grand Carénage）的一部分，法国正在对所有反应堆进行升级改造，尽可能接近第三代反应堆的安全要求。中国政府也采取了提高核电厂安全性的措施，防止严重事故情况下大量释放放射性物质（见附录 5 - 2）。

### 5.1.2 风险指引的纵深防御体系

建议"实际消除"（WENRA 的表述）任何可能导致大量或早期放射性释放的情景，通过提高安全裕量、采取补充安全措施和加强纵深防御等手段，大幅度降低剩余风险。

补充安全措施的设计和设置的主要原则是，核安全可合理达到尽量高的水平以及没有负面效应。为此，可以综合考虑各类因素，包括发生概率和剩余风险的后果，避免对正常运行、预计运行事件（anticipated operational occurrences，AOO）、DBA 和设计扩展工况（design extension conditions，DEC）的响应功能造成不利影响。

风险指引的纵深防御体系（risk-informed defence in depth system，RDIDS）如表 5-1 所示。该体系采用了专设安全设施、附加安全设施和补充安全措施：

在第 3 层次，专设安全设施用于应对 DBA，应按照安全级系统设备要求设计和建造；

第 4 层次引入附加安全设施用于应对 DEC，例如，在卸压系统安装快速卸压阀，防止高压熔堆（high pressure core melt，HPCM）以及可能由此导致的灾难性安全壳早期失效。从风险指引的角度考虑，不要求附加安全设施达到安全级（冗余、设备鉴定等）。例如，可以考虑在第 4 层次用消防系统补充乏燃料池，尽管该系统不是安全级。在 DEC 中可针对系统可靠性开展概率性评估，用于支持安全全过程系统分析。

在第 5 层次，使用补充安全措施以避免或减少极端工况下的剩余风险。补充安全措施包括安全壳过滤排放措施、场外应急计划、缓解核电厂大范围损伤后果的移动电源、移动泵、贮水池，以及由核电集团和国家层面配置的用于支援核电厂内外应急响应的移动设备等。原则上，补充安全措施不要求是安全级，只要求证明具备可靠性，并定期检查有效性。

表 5-1　风险指引的纵深防御体系

| 纵深防御层次 | 目标 | 基本措施 | 对应核电厂工况 |
|---|---|---|---|
| 第 1 层次 | 预防异常运行和失效 | 保守设计和高质量建造与运行 | 正常运行 |
| 第 2 层次 | 控制异常运行并检测失效 | 控制、限制和保护系统及监测设施 | AOO |
| 第 3 层次 | 将事故控制在设计基准以内 | 专设安全设施和事故响应程序 | DBA（假设单一始发事件） |
| 第 4 层次 | 控制严重工况，包括严重事故预防（4a）和后果缓解（4b） | 附加安全设施和事故管理 | DEC，包括多重失效（4a）和严重事故（4b） |
| 第 5 层次 | 极端工况下的工程抢险；放射性释放后果的缓解 | 补充安全措施、大范围损伤情况下的管理准则和场外应急响应 | 剩余风险 |

在风险指引的纵深防御体系框架下，第 4 层次要求在核电厂设计中包含应对 DEC 的附加措施，考虑其适当性和可靠性，在事故预防和缓解的设计措施之间达成更合理的平衡。相关核电厂安全分析应证明，在严重事故工况下，安全壳能维持完整性，不会发生向环境大量放射性释放的情况。根据对核电厂进行逐一分析的结果，如果无法证明安全壳的完整性，则应安装安全壳过滤排放系统。

在第 5 层次，我们假定第 4 层次纵深防御失效，尽管目标是实际消除大量放射性释放，但这种释放仍然发生了。因此，仍然有必要对可能出现的紧急状态有所准备（开展场外应急准备以减轻后果）。

### 5.1.3 新的安全威胁因素

在评估在运核电厂和新建核电厂的核安全时，需要特别考虑网络攻击和恐怖主义等新的安全威胁因素。

网络攻击并非是核电厂特有的威胁，因此，核电厂的防御级别应当与其他提供重要服务或可能会对环境造成潜在影响的大型设施相同。核电厂各个阶段（设计、建造、运行和维护）都在迅速实现数字化，因此应特别注意对各阶段的软件及数据库实施保护。

运营单位应该设置 CSO 岗位，并在 CSO 的带领下组建一个专门的网络安全部门，负责制定和实施组织内部各层级的网络安全政策[48]。网络安全部门还应负责审查分包商在此领域的合规情况。应特别关注仪表及控制（instrumentation and control，I&C）系统。鉴于如今的核电厂已实现 I&C 数字化，因此 I&C 的安全十分重要。I&C 系统是确保核设施安全的关键，可在必要时实施安全停堆，为此，我们应特别重视 I&C 系统的保护工作。这一系统不得接入外部网络，在变更和更新时应严格遵守相应程序、控制和再鉴定的要求。

虽然恐怖主义也并非是核电厂特有的威胁，但不幸的是，其导致的后果将相当严重。自民用核工业诞生以来，IAEA 已组织开展了大量工作，防止出现核材料的不可控扩散及使用[49]。多年实践证明，该防御体系非常有效并理应获得坚定支持。然而，我们必须重新考虑恐怖分子对核设施的直接攻击；“9・11”恐怖袭击事件表明，现代世界在新形式的恐怖主义面前十分脆弱。这一话题本质上是保密的，基本上不可能公开讨论各国预案。原则上，在应对这一特定威胁时同样应采用纵深防御理念。在设计（如抗震能力、安全壳承受严重超压的牢固性）方面使核设施具备抵抗某些外部入侵力量的能力，但可能需要附加的工程措施以保护构筑物的安全，并抵御飞机坠毁引起的高频振动。更重要的是，需要指定一个国家机构负责确定需要考虑的安全威胁因素；运营单位应与国家安全部队（警察、军队等）合作，共同制定这些威胁因素的防御应对措施。虽然无法提供细节信息，但应当说明工作理念，并应向经批准的议会成员或监管机构提供具体信息。

## 5.2 核电厂选址

核电厂的选址既要考虑到电力需求和电源布局要求，更要从安全角度考虑核电厂址的适宜性要求。国际业界关于核设施选址的基本要求达成了以下共识：①厂址安全；②环境保护；③应急准备。可以看出，应急准备仍是风险指引的纵深防御体系中的重要因素。

首先，应考虑以下 3 个方面：

（1）厂址所在区域发生外部事件的影响（这些事件可能是由自然或人为因素诱发）；

（2）可能影响放射性物质向人类和环境迁移的厂址及其环境特征；

（3）可能影响实施应急准备和响应的厂址因素。

在厂址安全评估方面，通常需要考虑以下 8 个评估要素：①地质和地震；②大气弥散；③非居住区和规划限制区；④人口分布；⑤应急计划；⑥安保大纲；⑦水文；⑧工业、军事和交通设施。

如果一个厂址经采用上述标准评估后被认为不适宜，且其缺陷不能通过设计、厂址保护措施或行政管理程序弥补，则必须排除该厂址，无需进一步考虑[50,51]。

为了避免外部事件造成的安全影响，核电厂选址时要深入考虑地质因素，避开地震断裂带、滑坡、火山等地质不稳定地区；还应考察气候、水文等因素，避免台风、海啸、潮汐、洪水等对核电厂带来威胁；同时，还要确保核电厂有足够的散热能力来排出余热。

此外，厂址选择时还需要考虑大件设备的运输、经济性以及公众接受度等方面的问题，尽管这些问题与安全无关。

内陆厂址和沿海厂址的安全要求是一致的，但考虑的因素（如台风、海啸、溃坝）可能有差异。内陆核电厂所面临的极端自然灾害情景可能包括：地震与滑坡、地裂/断层、塌陷；洪水与溃坝；地震与溃坝。

关于内陆核电厂发生事故后如何防止放射性废水进入河流的问题，中国开展了大量研究，制订出了事故后安全壳内废水处理的四项原则。确保放射性废水“能存储”“能封堵”“能处理”“能隔离”的四项原则可作为核电厂安全设计的补充安全措施，用于巩固核电厂的纵深防御体系，进一步保障核电安全。

法国也进行了类似的研发工作，提出了适用于各厂址和设施的解决方案，并定期进行审查。法国和中国负责此类事务的机构之间应积极开展交流。

在福岛第一核电厂事故发生后，中国核电发展受阻，尤其是内陆核电。由于“好的”沿海厂址短缺，一些以往被视为条件“不是很好的”沿海厂址（特别是地震风险比较高的厂址）被再次评估，并被认为适宜建设第三代核电厂。在地震风险较大的区域建造核电厂时，需要各方高度关注，进行包括安全裕量在内的深入研究，保守决策，确保安全。

法国每十年对核电厂址的适宜性进行一次审查，审查通过后方可批准在未来十年继续运营。数个厂址（以 Cadarache、Fessenheim 为例）设施寿期内的抗震设计标准有所提高；与此同时，还要证明设计有足够的裕量来应对更高的要求，而不会损害设施安全。

## 5.3 安全责任及政府职责

### 5.3.1 运营单位的主要责任

如果没有一个明确的组织机构负责保证设施的安全性并提供足够的资源履行职责，就没有安全可言。“安全的首要责任必须由对产生辐射危险的设施和活动负责的个人或组织承担”[52]。从法律角度看，负责人或组织是指核安全许可证持有者；其可以通过合同约定，将部分或全部运行维护工作委托他人；但核安全许可证持有者（有时称为“所有者或运营单位”）必须承担全部安全责任，并应证明有足够的资源来确保自己能够履行职责。

由于财务方面的约束，未来核电厂将具有复杂的所有权结构，这可能导致一个设施出现多个所有者，其可通过签订合同，将运营工作委托给其中一个所有者。在这种情况下，应当明确安全责任。核安全监管机构必须确保所有者或运营单位的组织结构清晰明确，这应是颁发许可证的条件之一。

为了履行安全责任，运营单位应拥有足够的技术和财务资源，以便能够开展和管理核安全活动。这些活动可能由内部工作人员执行，或是分包他人。为了控制和监督安全相关活动，通常会建立独立于运维部门的安全部门或安全处室。对于复杂的组织

（例如有多个场址、一个场址有多个机组），建议安全部门或安全处室设立双重报告渠道，即在业务运营方面向所处级别（机组、厂址、公司）的业务管理部门报告，并在职能上向上一级别的安全组织报告。此外，需要让每个员工或分包商有权通过组织内明确宣告的、独立于管理层的举报渠道秘密报告其目睹或发现的安全违规行为，并且举报人不存在任何受制裁风险（吹哨人制度）。建议大型组织设立独立检查部门，其不仅能够向组织的最高管理层进行报告，对系统进行审计，还能对设施进行自检；该部门不完全依赖于核安全监管部门，能够自行开展定期检查。

无论何种安全组织，均必须强调基于“安全第一”原则的安全文化，并在组织及分包商内部各个级别（从高级管理层到普通员工）之间传播。

IAEA 指出，核设施运营单位对核安全负首要责任，并建议：有必要与世界核电运营者协会（WANO）合作，获取最佳实践建议，帮助运营单位充分履行职责。

### 5.3.2　政府和监管机构的职责和作用

政府的职责是保护人员和环境。政府应围绕安全问题建立法律和政府框架，包括设立独立的核安全监管机构。继而，核安全监管机构根据核法规颁发建造和运行许可证。为了检查是否符合许可证的要求，核安全监管机构应对运营单位或许可证持有者进行监督检查。

但是，无论监管机构如何管控，这种监督检查不会减轻运营单位对核安全承担的全部责任。

为了实施上述原则，中国于 2015 年 7 月 1 日颁布了《中华人民共和国国家安全法》，将核安全与政治安全、国土安全、军事安全、经济安全、文化安全、社会安全、科技安全、信息安全、生态安全、资源安全一起纳入国家安全体系；并在 2018 年 1 月 1 日开始施行的《中华人民共和国核安全法》中，明确了各方的核安全责任。在法国，上述原则已被纳入《环境法典》（第 L591-1 条及其后各条）和相应法令中。

## 5.4　核安全与公众接受度

由于核电技术的复杂性以及诸如福岛第一核电厂事故等重大事故带来的负面影

响，公众仍然有“恐核”心理，对核能的和平利用心存疑虑。公众对核电的“邻避效应”（not-in-my-back-yard，NIMBY）日益突出，表现为对核电建设项目的抵制和反对情绪不断升级。无论核电在成本和二氧化碳排放方面有何优势，公众接受度已经成为制约核电发展的瓶颈。核安全问题的公众沟通工作任重道远。

改进核安全，更好地预防严重核事故和减轻其后果是进一步接受核能的先决条件。但同样重要的是，公众要意识到并了解这些改进。加强公众沟通，增强公众对核能的信心，是核能健康发展的重要组成部分。良好的公众沟通需要有效和透明的信息、积极的公众参与，以及与地方当局和公众的长期对话。在技术问题上，应当将更好地为公众提供技术教育作为教育体系的目标，这需要从教师和教育工作者开始，以及从小学开始。

核安全监管机构应注重实施公开透明的核安全监督和管理，构建“中央督导、地方主导、企业作为、社会参与”的公众沟通机制。虽然促进核能发展不是核安全监管机构的职责，但应向公众解释其履行职责的方式，以及为何有信心颁发核安全许可证。应完善政府网站，如搭建信息公开平台，以积极推进涉核项目环评报告、全国辐射监测结果、项目审批信息等相关文件公开。在政策制定和项目环评过程中应广泛听取公众意见，提高公众参与程度。

经验表明，信息公开是基础，公众参与是前提，利益共享是关键。若没有利益共享，即使公众对于核电风险意识有所增强且认知有所改观，仍然难以解决“邻避效应”问题。

总体而言，公众对接受现有厂址的扩建基本没有问题，这可能是因为当地公众（包括当地政府）已经比较了解核电，感受到核电带来的地区经济发展和社会发展的益处，而没有感受到与核电厂为邻带来任何安全风险。然而，新建厂址仍然面临着公众接受度问题，因为当地居民对于这类新项目尚未获得良好的体验。

## 5.5 结论

如第 4 章所述，第三代反应堆具有应对严重事故的专用预防和缓解措施，可以实现环境风险可控，完全能够满足核安全法规的要求。然而，需要说明的是，核安全是一个通过良好的经验反馈系统进行持续学习、更新和改进的领域。

通过深入分析国内外发生的各类核事故，以及借鉴其他行业面临风险时的最佳实践，核设施安全性已得到了有效改善。在分析了以往事故的根本原因并采取适当措施后，可以大幅降低潜在安全风险。自福岛第一核电厂事故后，核能的公众接受问题变得更加重要，甚至已经成为制约核能发展的瓶颈。因此，有必要说明核工业已采取了多重额外的安全措施来降低风险，在发生严重事故时能够消除大量放射性释放并保护人类和环境。

## 附录5－1　核安全监管体系

核能的开发利用给人类发展带来了新的动力，同时，核能发展也伴随着安全风险和挑战。由于核事故的后果不限于一国一地，因此必须认识到核能的跨国性质，并进行适当的国际合作。

核安全监管体系就像一座大厦，需要系统性地构建基础和支撑。典型的监管体系（即核与辐射安全监管大厦）如图A5－1所示，由四块基石、八项支撑组成，又称“四梁八柱”。

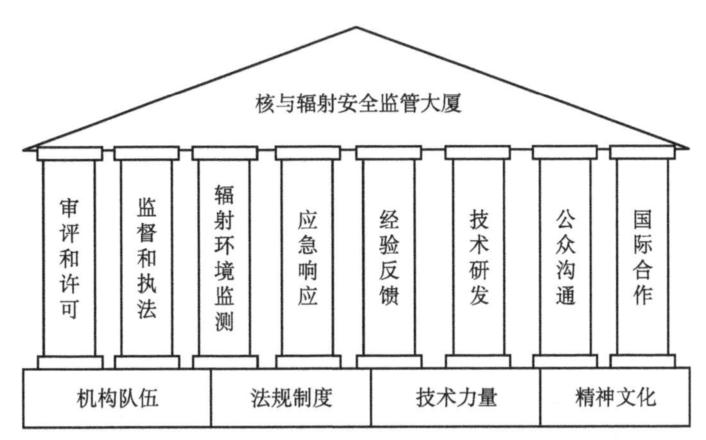


图A5－1　核与辐射安全监管大厦示意图

机构队伍、法规制度、技术力量、精神文化是核与辐射安全监管大厦的四块基石。一般认为，有必要按照以下建议来夯实四块基石。

（1）机构队伍：建立独立于核能开发部门的核安全监管机构；

（2）法规制度：完善以《原子能法》和/或《核安全法》为统领的核领域法律顶层设计；

（3）技术力量：构建独立分析和试验验证、信息共享、交流培训等平台；

（4）精神文化：普及核安全文化，强化风险意识，坚持“安全第一、质量第一”原则。

八个支柱领域包括审评和许可、监督和执法、辐射环境监测、应急响应、经验反馈、技术研发、公众沟通以及国际合作。在“四块基石”“八项支撑”的基础上构建用于核安全监管的健全有效的管理体系。

## 附录 5－2　中国采取的行动

2015 年 7 月 1 日，中国政府颁布《中华人民共和国国家安全法》，将核安全纳入国家安全体系。

此外，中国在 2012 年发布的《核安全与放射性污染防治“十二五”规划及 2020 年远景目标》中明确要求：“十三五”期间及以后新建核电机组力争实现从设计上实际消除大量放射性物质释放的可能性。2017 年发布的《核安全与放射性污染防治“十三五”规划及 2025 年远景目标》中明确提出：新建核电机组保持国际先进水平，从设计上实际消除大量放射性物质释放。

中国国家核安全局于 2016 年 10 月发布了新版的《核动力厂设计安全规定》（HAF102－2016）。作为中国核安全监管/法律体系中的重要文件之一，HAF102－2016 提出了与核动力厂安全重要的构筑物、系统和部件的设计、规程和组织流程所必须满足的要求，以及进行全面安全评估的要求。

HAF102－2016 在参照 IAEA《核动力厂安全：设计》（SSR2/1，第 1 版）的基础上，还吸收了美国核管会（NRC）、WENRA 等监管机构和组织发布的相关要求，如抵御商用飞机恶意撞击的要求。

“十二五”和“十三五”规划规定：

（1）在 DBA 和/或 DEC 工况下，核电厂事故不会导致放射性物质显著外泄。

（2）在极端工况下，避免发生大规模的放射性物质释放，以保护人员、社会和环境免受危害，特别是避免出现造成周围环境长期严重污染的类似于福岛第一核电厂事故的情景。“实际消除大量放射性释放”这一安全目标并不是要取消场外应急计划，因为福岛第一核电厂事故已经证明了场外应急响应的重要性。这里的“大量放射性释放”，是指类似于福岛第一核电厂事故时的放射性释放情景。

HAF102－2016 体现了三方面并重，即

（1）内部事件与外部事件设防并重；

（2）严重事故预防和缓解并重；

（3）确定论和概率论并重。

HAF102－2016 要求的主要改进包括：

（1）强化对于可导致公众和环境不可接受放射性后果的预防；

（2）采取严重事故缓解措施，避免早期释放和对周围环境的长期污染；

（3）强化核电厂设计，包括强化纵深防御第四层次以及考虑外部事件灾害并维持足够裕量等手段来预防严重事故；

（4）强化最终排热手段的可靠性；

（5）强化应急动力供应；

（6）增强燃料贮存的安全性，避免燃料裸露；

（7）提供接口，便于在必要时使用移动设备；

（8）强化应急响应设施的性能。

# 第6章　结　　论

本报告延续了之前中法三院相关专家的研究内容。

2017年由中法三院联合发表的第一期报告主要对核能未来发展提出了意见和建议。本报告则更具体地讨论了从铀矿开采到放射性废物处理的整个过程中核能对环境的影响，尤其重点讨论了与核能发电有关的四大环境问题，这些问题是：

- 核设施正常运行时放射性流出物释放的评估和控制；
- 乏燃料和放射性废物的长期管理，特别是要在地质处置库内处置的乏燃料和放射性废物；
- 严重核事故及其放射性物质释放的管理；
- 改进核安全可以限制其环境影响，有助于公众接受核能。

一方面，核电具备诸多优势，尤其是可以提供可调峰的电力来源，同时产生极少的温室气体。在全球变暖的背景下，核电几乎不释放温室气体，这使其成为一种独特的电力来源。此外，化石燃料发电厂释放大量空气污染物，例如颗粒物、氮氧化物、硫氧化物和重金属，而核电则不会产生空气污染物。另一个优势是，核电需要的土地面积相对较少。核能生产具有足够的灵活性，可以填补大部分的间歇性可再生能源。

总结上述优势，可以得出以下结论：核能是解决环境变化问题最有效的方案之一，也是能源转型中最适当的能源之一。没有核能，很难实现温室气体减排的目标。

但另一方面，也需要评估核能可能对环境造成的不利影响，此即本报告的重点。

首先需要指出，在正常条件下，核能对环境的影响有着良好的记录，而且环境中放射性核素的浓度是很容易测量的，因此可对此类设施进行独立监控。核能国家都会根据辐射防护的安全规定对流出物的辐射水平进行管控，实际的释放量仅为授权水平的百分之几，而授权水平本身也远低于天然辐射的影响，这就是为什么本报告能够得出如下结论：在正常运行条件下，核电厂的辐射影响是微乎其微的，或者说是非常有限的。

报告随后考虑了冷却水的问题。核电厂通常建造在海岸附近，利用海水来解决冷却问题。这种海水在被排放到海里之前，在热交换器装置中温度会稍微升高，而不会产生任何影响。

在另外一些情况下，核电厂也会建造在大型河流附近，冷凝器通过一次循环冷却（冷却水返回河流），或经过冷却塔冷却。对于第一种方案，由于核电厂下游会产生“热污染”（水温升高），因此内陆核电厂的运营可能面临更多限制。冷却塔大大降低了对河流的热影响，但不利于取水。在大型河流沿岸兴建新核电厂时，应当更好地向公众公开为控制水温及限制取水而采取的措施。

放射性废物管理的每一步均需考虑环境保护问题：

- 废物包的隔离和密封；
- 针对各类放射性废物的近地表或深地质设施的贮存和处置方法。

尖端工程技术的发展、放射性核素/有毒物质在工程屏障和岩石圈中行为的研发成果以及大规模的国际合作，都将促进放射性废物管理方案的进步。

从放射性废物的产生到处置库的处置（将放射性废物与岩石圈隔离）期间，所有的操作都会得到监测，其中处置库可以将废物包装与生物圈隔离开来。这些设施周围的本底水平将永久性地被监测，反馈的监测结果表明正常的释放值远远小于设施许可时安全和环境管理机构的预测值和批准值。

处置库关闭后，监控将在测试期间持续进行，安全性将从能动变为非能动。大多数放射性核素将在贮存库中衰变，可能返回生物圈的放射性核素的衰变时间会非常长，长到其放射性毒素的影响可以忽略不计。尽管分析实验室和地下岩石实验室获取的可用数据是短期数据，但是自然类似物可以为废物处置库的建模和安全评估提供宝贵支持。例如，加蓬 Oklo 的天然核反应堆可以在数百万年的时间里控制锕系元素和裂变产物的释放，地中海考古发现的玻璃可以在数千年的时间里抵制风蚀和雨磨。

报告讨论的一个主要问题是严重事故对环境的影响，并对这些事故所引起的问题进行了全面、透明、独立和平衡的评估，其中这些严重事故塑造了核能发展的历史。关于核反应堆的三大事故（三里岛、切尔诺贝利、福岛），许多文献已经作出了详细的阐述。关于燃料循环设施的几次严重事故，文献的阐述却不够翔实。因此，我们需要采取行动，对后一类事故进行研究和阐述。本报告认为：一方面，7 级 INES 事故（切尔诺贝利和福岛）对环境造成了重大影响，降低了公众对核能发电系统的信心。另一方面，从事故中获得的经验反馈则在诸多方面都大大提高了核电的安全性，比如反应堆设计、运行管理和重大事故管理指南。事实证明，这是相当有价值的。

将来可能发生的重大事故对环境的影响已经大大降低。此外，在运和在建的核电厂都制定了减轻损失的措施和方法，这将减少事故对环境的影响。这些措施的目的是在最大程度上减少受影响的区域，避免污染和大规模人员的长期撤离。

一个尚未解决的问题是低剂量或极低剂量的长期效应。尽管世界上绝大多数流行病学研究证明它们是无害的，但是科学界和核能界还未就此问题达成一致。

严重事故的全面保护和缓解措施提高了第三代反应堆的安全水平。第三代反应堆配备了先进的大型安全壳，能够抵御外部危害并保持完整性，从而避免放射物质释放至周围环境。

经验反馈推动了现有核电厂的升级，改进了新反应堆设计，也改进了目前由核电厂营运单位实施的安全指南，从而大大降低了切尔诺贝利和福岛等事故再次发生的可能性。如果发生这样的事故，放射性物质的泄漏将被降到最低，并且不需要大规模或长期的人员疏散。如果国际原子能机构（IAEA）或世界核电运营者协会（WANO）的全球评估能够证明全球范围内在运的核电厂已经进行了高水平升级，那么这将是非常有意义的。

鉴于安全管理对环境保护至关重要，报告强调了以下观点：

● 事实证明，风险导向型的纵深防御体系是一种更加先进的、完整的安全方法。它由 5 个安全等级构成，可以大大降低严重事故的剩余风险和发生概率，这相应地也会对环境产生重要影响。

● 核电厂的选址不仅应当考虑电力需求和电厂布局，还应从安全的角度出发，考虑选址的适宜性，也就是综合衡量选址的安全性、环境保护和应急准备状态。另外，核设施选址的基本要求应在国际范围内达成共识。

● 安全机构在安全改进和控制方面发挥着重要的作用，但核电厂运行方仍要承担全部责任。双方应积极开展对话，提供最高水平的环境保护。

总的来说，本报告的目的是全面评估核能对环境的影响。一方面，核能在生产能源方面具有诸多优势，且排放的温室气体极少，不像化石燃料能源系统那样排放空气污染物、固体纳米或微小颗粒。当前，人类活动导致的气候变化已成为人类面临的最困难的挑战之一，空气污染也成为许多国家面临的主要问题，在这种情况下，核能具有不可或缺的价值。另一方面，放射性废物管理和严重事故的多重后果，也使人们更加关注核能对地区乃至全球环境的影响。人类已经做出了相当大的努力，以制定出可持续管理高放废物的方案，从而使其在地质结构中得到最终处置。从三大事故中吸取的经验教训有助于优化核反应堆设计并降低放射性物质释放发生的概率，并可以确保即使发生事故核设施对环境的影响依然有限。

# 参考文献


[1] International Energy Agency. World Energy Outlook 2018.

[2] http://www. french-nuclear-safety. fr/Information/Publications/ASN-s-annual-reports/ASN-Report-on-the-state-of-nuclear-safety-and-radiation-protection-in-France-in-2017：144，145.

[3] NCRP. Ionizing Radiation Exposure of the Population of the United States. NCRP Report No. 160，2009.

[4] 潘自强，刘森林．中国辐射水平．北京：中国原子能出版社，2010.

[5] UNEP. Radiation Effects and Sources. 2016：29，54.

[6] Brookins G，Douglas. Migration and retention of elements at the Oklo natural reactor. Environmental Geology，1982，4：201－208. 10. 1007/BF02380513.

[7] http：//www. french-nuclear-safety. fr/Information/Publications/ASN-s-annual-reports/ASN-Report-on-the-state-of-nuclear-safety-and-radiation-protection-in-France-in-2017：53.

[8] "核电链和煤电链排放的放射性影响评价"项目组．不同发电能源排放的放射性影响评价．2017.

[9] WHO，World Health Organization. Guidelines for Drinking-Water Quality. Vol. 1. Third Edition. Geneva，Switzerland，2004.

[10] Le tritium dans l'environnement-Synthèse des connaissances IRSN Rapport DEI 2009－05-https：//www. irsn. fr/FR/expertise/rapports_expertise/Documents/environnement/IRSN_DEI-2009－05_Tritium-environnement-synthese-connaissances. pdf.

[11] Actualisation des connaissances acquises sur le tritium dans l'environnement-IRSN-https：//www. irsn. fr/FR/expertise/rapports_expertise/surveillance-environnement/Documents/IRSN_Rapport-Tritium-2017_PRP-ENV-SERIS-2017－00004. pdf.

[12] Tritium and the environment-IRSN-August 2012-July 2017-https：//www. irsn. fr/EN/Research/publications-documentation/radionuclides-sheets/environment/Pages/Tritium-environment. aspx.

[13] 2014：Annex III：Technology-specific cost and performance parameters. In：Climate Change 2014：Mitigation of Climate Change. https：//www. ipcc. ch/site/assets/uploads/2018/02/ipcc_wg3_ar5_annex-iii. pdf page 1335.

[14] Global land outlook working paper energy and land use. https：//global-land-outlook. squarespace. com/s/Fritsche-et-al-2017-Energy-and-Land-Use-GLO-paper-corr. pdf.

[15] DoE quadrennial technology 2015. https：//www. energy. gov/sites/prod/files/2017/03/f34/qtr-2015-chapter10. pdf see note 47 page 390 and 410 for reference publications.

[16] Department of Energy. Quadrennial technology review. 2015 Sep. https：//www. energy. gov/sites/prod/files/2017/03/f34/qtr-2015-chapter10. pdf. p. 390，410.

[17] Centrales nucléaires et environnement-Prélèvements d'eau et rejets-EDP Sciences Page 169. Jun 2014. https：//www. edp-open. org/books/edp-open-books/278-centrales-nucleaires-et-environnement-prelevements-deau-et-rejets.

[18] Life cycle water use for electricity generation：a review and harmonization of literature estimates. Published 12 March，2013. https：//doi. org/10. 1088/1748-9326/8/1/015031.

[19] Revue Générale Nucléaire. SFEN，08/01/2019.

[20] Synthèse de l'étude thermique du Rhône-Agence de l'eau Rhône-Méditerranée-Corse，ONEMA，Agence Régionale de Santé Rhône-Alpes，Agence de Sûreté Nucléaire，CNR and EDF. May，2016.

[21] Le Quéré C，et al. The global carbon budget 1959-2011-Earth System Science Data Discussions 5，2013，2：1107-1157.

[22] IEA Statistics © OECD/IEA 2014.

[23] IEA Energy Technology Perspectives 2017.

[24] Nuclear Energy Agency-OECD-Accelerator-Driven Systems (ADS) and Fast

Reactors (FR) in Advanced Nuclear Fuel Cycles-2002.

[25] Assessment of the environmental footprint of nuclear energy systems. Comparison between closed and open fuel cycles. -Ch. Poinssot, S. Bourg, N. Ouvrier, N. Combernoux, C. Rostaing. Elsevier Energy, 2014.

[26]“不同发电能源温室气体排放关键问题研究”项目组．中国不同发电能源的温室气体排放．北京：中国原子能出版社，2015.

[27] Les études épidémiologiques des leucémies autour des installations nucléaires chez l'enfant et le jeune adulte: revue critique-Dominique LAURIER, Marie-Odile BERNIER, Sophie JACOB, Klervi LEURAUD, Camille METZ, Éric SAMSON, Patrick LALOI-IRSN-2008-https: //www. irsn. fr/FR/Larecherche/publications-documentation/aktis-lettre-dossiers-thematiques/RST/RST-2008/Documents/CHO3-4-Epidemio-Leuc. pdf.

[28] Spix C, Schmiedel S, Kaatsch P, et al. Case-control study on childhood cancer in the vicinity of NPPs in Germany 1980-2003. European Journal of Cancer, 2008, 44 (2): 275-284.

[29] 放射性废物分类. 2017，中国.

[30] 乏燃料管理安全和放射性废物管理安全联合公约第四次履约国家报告. 2017，中国，第 D. 4 节.

[31] 高放废物地质处置研究开发规划指南. 2006，中国.

[32] 中国工程院重点咨询项目研究报告：中国放射性废物管理战略研究. 2018.

[33] Assessment of the anticipated environmental footprint of future nuclear energy systems. Evidence of the beneficial effect of extensive recycling. J. Serp, Ch. Poinssot, S. Bourg. Energies, 2017.

[34] Report of the President's Commission on the Accident at Three Mile Island. Washington, D. C. , 1979.

[35] IAEA. INSAG-1: Summary Report on the Post-Accident Review Meeting on the Chernobyl Accident. 1986.

[36] IAEA. Environmental Consequences of the Chernobyl Accident and Their Remediation: Twenty Years of Experience. 2006.

[37] IAEA, WHO, UNDP, et al. Chernobyl's Legacy: Health, Environmental and Socio-Economic Impacts and Recommendations to the Governments of Belarus, the Russian Federation and Ukraine, Second revised version. 2006.

[38] UNSCEAR. Health effects due to radiation from the Chernobyl accident. 2011.

[39] 潘自强．切尔诺贝利和福岛事故对人体健康影响究竟有多大．中国核电，2018，11 (1)：11-14.

[40] International Atomic Energy Agency. The Fukushima Daiichi Accident-Report by the Director General. 2015.

[41] The National Diet of JAPAN, Fukushima Nuclear Accident Independent Investigation Commission. The Official Report of the Fukushima Nuclear Accident Independent Investigation Commission. 2012.

[42] IAEA. Fundamental safety principles: safety fundamentals. Chapter 2. Vienna, 2006.

[43] Electricité de France. Annual Public Information Report for the PENLY NPP. Paris, 2017.

[44] USNRC. Reactor Safety Study: An Assessment of Accident Risks in U. S. Commercial Nuclear Power Plants. Executive Summary. WASH-1400 (NUREG-75/014). Washington, 1975.

[45] USNRC. Severe Accident Risks: An Assessment for Five U. S. Nuclear Power Plants. NUREG-1150. Washington, 1990.

[46] WENRA. Safety of New NPP Designs. Study by Reactor Harmonization Working Group RHWG. March 2013.

[47] WENRA. Safety Reference Levels for Existing Reactors. Update in Reation to Lessons Learned from Tepco Fukushima Daiichi Accident. 2014.

[48] IAEA. Computer Security at Nuclear Facilities. Nuclear Security Series No. 17. Vienna, 2011.

[49] IAEA. The International Legal Framework for Nuclear Security. International Law Series No. 4. Vienna, 2011.

[50] IAEA. Site Survey and Site Selection for Nuclear Installations. No. SSG-35.

Vienna, 2015.

[51] IAEA. Managing Siting Activities for Nuclear Power Plants. No. NG-T-3. 7. Vienna, 2012.

[52] IAEA. Fundamental Safety Principles: Safety Fundamentals. Principles 1 and 2. Vienna, 2006.

# 词 汇 表

ADS：accelerator driven system，加速器驱动系统

AI：artificial intelligence，人工智能

ANCCLI：Association Nationale des CLIs，法国全国性 CLI 协会

Andra：Agence National pour la Gestion des Déchets Radioactifs，法国放射性废物管理局

AOO：anticipated operational occurrences，预计运行事件

ASN：Autorité de Sûreté Nucléaire，法国核安全局

ATF：accident tolerant fuel，耐事故燃料

BAT：best available technology，最佳现有技术

BDBA：beyond design basis accident，超设计基准事故

BWR：boiling water reactor，沸水堆

CA：control area，对照地区

CAE：Chinese Academy of Engineering，中国工程院

CAEA：Chinese Atomic Energy Authority，中国国家原子能机构

CCGT：combined cycle gas turbine，联合循环燃气轮机发电厂

CCS：carbon capture and storage，碳捕获和封存

CEA：Commissariat à l'Energie Atomique，法国原子能和替代能源委员会

CEFR：China Experimental Fast Reactor，中国实验快堆

CFC：close fuel cycle，闭式燃料循环

CIAE：China Institute of Atomic Energy，中国原子能科学研究院

Cigeo：Centre Industriel de Stockage Géologique，法国深层地质处置库

CLI：Commission Local d'Information，地方利益相关方信息委员会

CNE：Commission Nationale d'Evaluation，法国国家废物战略和研发评估委

员会

CNNC：China National Nuclear Corporation，中国核工业集团有限公司

CNPE：China Nuclear Power Engineering Corporation，中国核电工程有限公司

CSA：Centre de Stockage de l'Aube，法国低放废物的地表处置库

CSM：Centre de Stockage de la Manche，法国低放废物的地表处置库

CSO：chief security officer，首席安保主任

CSP：concentrating solar power，集中式太阳能热发电

DBA：design basis accident，设计基准事故

DCH：direct containment heating，安全壳直接加热

DEC：design extension conditions，设计扩展工况

DOE：Department of Energy，USA，美国能源部

EDF：Electricité de France，法国电力集团

EDMG：extensive damage mitigation guidelines，大范围损伤缓解导则

EPO：environmental permanent observatory，环境永久观测站

EPR：European Pressurized Water Reactor，欧洲压水堆

EPRI：Electric Power Research Institute，美国电力科学研究院

EU：European Union，欧盟

FARN：La Force d'Action Rapide Nucléaire，法国电力集团提出的“核能快速行动救援力量”

FR：fast neutron reactors，快中子反应堆

Gen-Ⅱ，Gen-Ⅲ，Gen-Ⅳ：目前正在运行或开发的第二代、第三代和第四代核反应堆（第一代是原型堆，现在已经退役）

GFR：gas cooled fast reactor，气冷快中子反应堆

GHG：greenhouse gas，温室气体

GIAG：Guide d'Intervention en Accident Grave，严重事故管理指南

GIF：Generation-Ⅳ International Forum，第四代核能系统国际论坛

GSG：general safety guide，通用安全指南

GWa/GWy：energy produced by one GW during one full year，一年生产 1 GW 能量

HBRA：high background radiation area，放射性高本底地区

HLW：high level waste，高放废物

HPCM：high pressure core melt，高压熔堆

HPR1000：advanced pressurized water reactor developed in China (also named Hualong One)，中国研制的先进压水堆“华龙一号”

HWR：heavy water reactor，重水反应堆

IAEA：International Atomic Energy Agency，国际原子能机构

I&C：instrumentation and control，仪表及控制系统

ICRP：International Commission on Radio Protection，国际辐射防护委员会

ICPE：Installation Classée Pour l'Environnement，法国环境保护分类设施（法国按照潜在环境污染风险对设施进行分类，指对环境保护重要的、最危险的设施）

IEA：International Energy Agency，国际能源署

IL-LLW：intermediate level-long lived waste，长寿命中放废物

INES：International Nuclear Event Scale，国际核事件分级表

IRSN：Institut de Radioprotection et Sûreté Nucléaire，法国国家辐射防护和核安全研究所

LBLOCA：large break loss of coolant accident，大破口失水事故

LCA：life cycle analyses，生命周期分析

LFR：lead-cooled fast reactor，铅冷快中子反应堆

LIL-SLW：low and intermediate level-short lived waste，短寿命低中放废物

LLW：low level waste，低放废物

LL-LLW：low level-long lived waste，长寿命低放废物

LOCA：loss of coolant accident，冷却剂丧失事故

LPCRP：*Law on Prevention and Control of Radioactive Pollution*，《放射性污染防治法》

MCCI：molten core concrete interaction，堆芯熔融物与混凝土相互作用

MEE：Ministry of Ecology and Environment，中国生态环境部

MOX：mixed uranium-plutonium oxide，铀钚混合氧化物

MSFR：fast spectrum molten salt reactor，快中子熔盐反应堆

MSR：molten salt reactor，熔盐反应堆

NIMBY：not-in-my-back-yard，邻避效应

NISA：Nuclear and Industrial Safety Agency（Japan），日本核能与工业安全局

NNSA：National Nuclear Safety Administration（China），中国国家核安全局

NPCSC：National People's Congress Standing Committee，全国人大常委会

NPP：nuclear power plant，核电厂

NRC：Nuclear Regulatory Commission（USA），美国核管理委员会

NRSC：Nuclear and Radiation Safety Center，生态环境部核与辐射安全中心

OECD：Organization for Economic Co-operation and Development，经济合作与发展组织

OFC：open fuel cycle，开式燃料循环

OTC：once through cycle，一次通过式燃料循环

PNGMDR：*Plan National de Gestion des Matières et Déchets Radioactifs*，《国家放射性材料与废物管理计划》

PRA：probabilistic risk assessment，概率风险评价技术

PRIS：Power Reactor Information System，国际原子能机构动力堆信息系统

PV：photovoltaics，光伏

PWR：pressurized water reactor，压水反应堆

R&D：research and development，研发

RDIDS：risk-informed defence in depth system，风险指引的纵深防御体系

SAMG：severe accident management guidelines，严重事故管理导则

SBLOCA：small break loss of coolant accident，小破口失水事故

SFR：sodium-cooled fast reactor，钠冷快中子反应堆

SGTR：steam generator tube rupture，蒸汽发生器传热管破裂

TBR：technical basis report，技术基础报告

THMC：thermal，hydrogeological，mechanical and chemical，热、水文地质、机械和化学

TMI：Three Mile Island，三里岛

TTC：twice through cycle，二次通过式燃料循环

UNEP：United Nations Environment Program，联合国环境规划署

UNSCEAR：United Nations Scientific Committee on the Effects of Atomic Radiation，联合国原子辐射效应科学委员会

$UO_x$：uranium oxide，氧化铀

URL：underground research laboratory，地下实验室

VLLW：very low-level waste，极低放废物

VVLLW：very，very low-level waste，非常、非常低水平的废物

WANO：World Association of Nuclear Operators，世界核营运者协会

WENRA：Western European Nuclear Regulators' Association，西欧核监管机构协会

WHO：World Health Organization，世界卫生组织

WOG：Westinghouse Owner Group，西屋公司所有者集团

# 后　　记

2017年年中，COP21和COP22承诺大幅减少全球温室气体排放。中法三院（中国工程院、法国国家技术院和法国科学院）全面研究了核能逐步取代化石燃料的可能性，概述了核能作为可靠和可调度的电力来源的优点，并在项目管理、教育和培训、研究和技术发展领域提出了建议，以进一步提高公众对这项技术的接受程度。

然而，核能对环境的潜在影响受到公众的强烈关注，中法三院决定在这第二份报告中特别指出这一点。本报告评估了核能的生命周期，包括铀矿开采、反应堆运行拆除、事故后果。

研究发现，包括燃料循环在内的核活动辐射影响只占自然辐射的一小部分，其中大部分来自铀矿开采，建议进一步减少这些影响。本报告将核能发电的环境影响与其他能源进行了比较，说明了核能在土地和材料方面的优点，特别强调了取水和消耗，这可能是一些内陆河流电厂面临的问题，但可以通过恰当设计冷却系统来缓解。

本报告全面讨论了废物管理，分析了基于最新废物封存技术的现行废物管理政策，包括在深地处置库中处置长寿命高放废物。安全分析结果表明，经过数千年或数百万年，放射性核素可能通过包括贮存库本身在内的限制屏障发生迁移，由此可能产生额外放射性，但这些额外放射性也将保持在天然本底辐射的1%以下。不过，废物管理仍有待进一步发展。

世界上几个地方（三里岛、切尔诺贝利、福岛）发生的几起堆芯熔化事故引发了公众担忧。本报告从这些事故中总结经验教训，提出了适用于新设施的安全要求，包括预防和减轻措施，以期大幅度减少核电厂边界以外的外部后果，避免长期疏散人口的需要。建议对现有设施进行升级，尽可能地满足这些要求。

本报告提出要改进安全分析，包括拓展传统的纵深防御，在反应堆的设计和运行中考虑严重的堆芯熔化事故，从而在最不可能的情况下满足“无疏散”目标。

本报告讨论了公众对核能的接受度，虽然这个问题因国家而异，但总的来说，向受过良好教育的公众提供透明的信息是至关重要的。

中法三院为独立机构，本报告仅代表中法三院立场，不代表核电领域工业参与者的立场，也不代表法国或中国政府的立场。

# Synthesis and recommendations

In August 2017, the three Academies produced a series of joint recommendations for the future of nuclear energy. The report was presented as a side event of the General Assembly of International Atomic Energy Agency (IAEA) held in Vienna, September 2017. This second report deals more specifically with the impacts of the nuclear energy cycle on the environment in response to a strong expectation of society for integrating environmental issues into all human activities. In this respect, the public expresses concerns about radioactive impact of nuclear power plants (NPPs) under normal operation or accidental conditions and for the long term about the return of radioactive elements to the biosphere from radwaste disposed of in geological layers.

The Academies have examined all operations from uranium mining to radioactive waste disposal and evaluated global and local as well as short-and long-term impacts in normal or accidental situations. The analysis considers consequences for human beings and ecosystems. It summarizes lessons learnt and actions that have already been or might be taken, to sustainably improve environmental protection. In this respect, it is concluded that the next Gen-Ⅲ NPPs and their associated facilities and the future Gen-Ⅳ NPPs will potentially feature a reduced environmental footprint.

Considerations and recommendations are synthesized in what follows. The Academies are aware that most of these recommendations correspond to actions which have been undertaken by the nuclear energy stakeholders. They wish to point out that these are valuable actions that should be pursued, and in some cases, that the effort should be intensified.

# 0.1 Impacts from NPPs and nuclear fuel cycle facilities under normal operation

## 0.1.1 Nuclear energy and global impacts on the environment

According to several life cycle analyses (LCA), nuclear energy generates low amounts of $CO_2$ per MWh. These emissions are as low as those of hydroelectric energy, notably better than photovoltaic and just a little higher than wind energy. It is however important to recall that intermittent energy sources need to be compensated when they are not available and that this modifies their environmental performance. In terms of consumption of standard materials for construction and critical metallic materials, nuclear energy requires much lesser amounts than photovoltaic and wind energy for the same energy production. Radiological impacts mainly originate from the releases of radioactive gases (rare gases, tritium, radon, and others) and liquid effluents (mainly tritium) to the environment. The radiological impact on the public is a very small fraction-less than about 1%-of the overall impact of natural sources of radiation. There is a debate about long-term effects of low and very low dose/dose rate exposures; however most epidemiological studies around the world provide no evidence of their effects on the life realm; molecular epidemiological studies taking into account identified effects at cell level could prove more efficient and should be encouraged. It is also worth noting that some species, like insects, may withstand high radiation levels.

## 0.1.2 Nuclear energy and local impacts on the environment

While fossil fuel plants (and in particular those using coal or lignite) emit large amounts of air pollutants such as particles, nitric oxides, sulfur oxides, heavy metals and various other releases of chemicals, this is not the case for NPPs. In this respect, fired coal plants release large quantities of natural radioactivity mainly in the form of gaseous radon and their solid waste contains sizable amounts of uranium and thorium that are managed as radioactive waste (TENORM-technologically enhanced naturally occurring radioactive material). Thus, nuclear energy has in fact positive effects regarding local area if it leads to the closure of fossil fuel plants. The absence of emissions brings a notable improvement in air quality and reduces damages to the environment such as those of acid rain. The main environmental footprints are those associated with front-end facilities. The front-end of the nuclear fuel cycle starts with ore extraction in the mines and ends with the delivery of the enriched uranium to the nuclear fuel assembly producer ($UO_x$ fuel; it includes handling of plutonium for MOX fuel fabrication).

Production of non-radioactive technical waste from the construction of reactors is lower than that associated with the construction of wind turbine or photovoltaic devices, as quantitatively analyzed in Chapter 2.

Land occupation with regard to energy produced is also significantly lower for nuclear energy than that needed for PV or wind farms. Around two thirds of land use is due to mining and decommissioned NPPs.

The withdrawal of water from rivers for cooling NPPs is of noteworthy importance, higher per MWh than fossil fuel facilities. The water stress and temperature increase need to be considered when siting inland NPPs in view of water availability. In general, most of the water is returned to the river but the present climate change already exhibits warm and dry episodes which occasionally force to operate below the nominal power. The potential impact of global warming should be carefully anticipated.

The Academies consider that a proper evaluation of the impacts of nuclear energy on the environment requires that:

- Exposures induced by nuclear activities should be compared in all cases with natural exposures.

- Background epidemiological studies should be carried out before any operation of a new nuclear facility. They are important and necessary for the comparative analysis of any post-accident epidemiological study, the analysis of radiation risks, and the responses to the public concerns.
- Water stress and future climate change should be considered for siting of inland NPPs.

In addition, the Academies make a general recommendation to reduce the footprint of nuclear energy by:

- Actively developing advanced nuclear technologies that canreduce the impacts on the environment from the front-end of the nuclear fuel cycle operations. The impact from nuclear front-end activities is higher than that from the back-end, defined as the management of the spent fuel up to geological disposal. Gen-Ⅳ NPPs, based on low consumption of uranium, will be beneficial to the environment when they become operational. Indeed, fast neutron reactors or multi-recycling spent fuel have the potential to drastically reduce these impacts. Preparation of their commercial development needs to be pursued.

In general, and except for water, nuclear energy uses low amounts of materials per installed MW, and its radiological and non-radiological impacts to the environment under normal operation and throughout the fuel cycle are limited.

## 0.2 Impacts from NPPs and nuclear fuel cycle facilities in accidental situations

The main environmental impacts of nuclear energy resulted from severe accidents (ranked at level 6 or 7 on the International Nuclear Event Scale, INES) that have marked the history of nuclear energy development. The accidents of Chernobyl NPP and Fukushima Daiichi NPPs have had a tremendous impact on the public opinion and on the development of nuclear energy worldwide. Lessons learned from these accidents and from the Three Mile Island accident (ranked at level 5 on INES) have led to major technology changes in reactor designs and operational procedures, which are implemented in Gen-Ⅲ NPPs and retrofitted in operating reactors as far as reasonable. Environmental risks in the event of a severe accident that might occur in the future have been substantially reduced so that they remain confined to the NPP site premises.

The Academies recommend:

● To continue research on the mechanisms leading to severe accidents (internal events like critical excursion, loss of cooling or external events like earthquakes, plane crash, terrorist attack) and provide support for their prevention and mitigation. Further studies to maintain the integrity of containment, or develop accident tolerant fuels (ATF) should also be pursued.

● To further accumulate experience in the implementation of severe accident management guidelines and to implement prevention and mitigation measures aimed at coping with large-scale damage in NPPs and multi-unit accidents, and to strengthen emergency response capabilities.

## 0.3 Impacts from radwaste management

Nuclear energy yields short-and long-lived radwaste. The management of the former is implemented through industrial channels leading to their disposal in near surface repositories. The management of the latter depends on their radiological activities. The most radioactive waste (spent fuel or high-medium level radwaste from reprocessing) is intended to be disposed of in deep geologic formations. Radwaste from mining/refining uranium is properly disposed of (for mining waste, mostly *in-situ*). The immediate impacts on the environment mainly originate from the releases of effluents from processing/packaging crude radwaste. Under the present practices, these activities have very low local and global impacts on health and the environment. According to many simulations supported by a large database, long-term deferred impacts, if any, are expected to be less than the impacts of natural radiation. Nevertheless, as perceived by the public, the management of radwaste is one of the major challenges of nuclear energy.

In order to improve the understanding of the real impactsof radwaste management on the environment, the Academies recommend that:

- The methodology to evaluate all environmental impacts (radiological and chemical) and the associated risks should be improved taking into considering the waste originating both from the front-and back-end of the nuclear fuel cycle, and time scales.

To support this general recommendation, the Academies propose that:

- Quantitative parameters should be defined to characterize the hazards linked to radwaste in order to better cope with environmental issues;
- R&D programmes should be further developed, which is aimed at a better understanding of the radiological and chemical impacts on ecosystems (reversibility, resilience, bio-availability of elements of interest …);
- A comprehensive and responsible system should be used to protect the environment (including legislation, competent and independent bodies, and funding processes), and be made clear and visible to the public;

In general, only the best available technologies (BAT, provided they are robust and with a high technology readiness level) can confine the radionuclides at every step of the processes.

## 0.4 Nuclear and radiation safety/security as a tool to prevent impacts on the environment

One main goal of nuclear safety is to eliminate the possibility of large radioactive releases from severe accidents into the environment; it is one major problem of nuclear energy. Nuclear and radiation safety, which is the responsibility of designers, operators and safety authorities, has a key role in environmental protection. The goal of security is to prevent malevolent action on nuclear facilities which could also lead to the release of radioactivity. Security is a governmental responsibility.

The Academies recommend that the owners of nuclear facilities:

- Test the resilience of the existing nuclear facilities to external events, which should be higher than considered in the design basis;
- Upgrade existing nuclear facilities to meet the same safety objectives as that set for new facilities, and be reasonably achievable;
- Implement the risk-oriented defence in depth, including beyond design basis conditions, for all facilities;
- Perform external additional reviews of their safety management systems, and not exclusively rely on the reviews carried out by the safety authorities.

As environmental protection is a major sensitive issue for people, it is recommended that nuclear regulatory agencies:

- Organize a transparent supervision of nuclear safety, and enforce transparent communication;
- Establish a permanent dialog with local authorities and the public.

The Academies consider that a collective effort should be made to educate and inform the public about nuclear energy matters in particular those related to the environmental impact.

With all these conditions being met, the Academies consider that the requirements to protect the environment are best implemented with energy mixes including nuclear energy in conjunction with renewable energies.

# Chapter 1　Introduction

On one hand, nuclear power boosts many advantages, in particular providing an on-demand dispatchable source of electrical energy with extremely low levels of air pollutant emissions and greenhouse gases (GHG). In the present context where climate change has become perhaps the most important problem, this characteristic is a fundamental asset of nuclear energy. The fact that it produces very limited air pollutants is also of importance when one considers the major degradation of air quality in many parts of the world. On the other hand, like all other sources of energy, the nuclear fuel cycle has environmental impacts. This report provides a comprehensive evaluation of these impacts and reviews how it is being limited and controlled; it is thus specifically focused on environmental and safety issues and does not attempt to cover any other topics.

Chapter 1 comprises three parts. The first part briefly considers trends in energy demand. The second part discusses decarbonization commitments and $CO_2$ emissions from various energy sources. The third part introduces the problems of environmental protection as a requirement to make nuclear energy sustainable.

## 1.1　Trends in energy demand

According to many national prospects dealing with the future energy mix, it appears that, energy demand will not increase globally on a short-term basis. But on a medium-and long-term basis, the increase of the world population, per capita revenues and the improvement of the quality of life will result in the demand increase in most countries. Thus, global energy demand will inevitably increase and that will mainly take the form of a growth in electricity demand (IEA, Ref. [1]). One way to provide massive electricity delivery while avoiding the combustion of fossil fuels is to use nuclear energy.

At present, sixteen countries make extensive use of nuclear energy which provides more than 20% of the electrical power supply in each of these countries (Figure 1-1). Of the 29 European countries (27 EU member countries+Switzerland+UK), fifteen have NPPs with a total of 132 units, delivering 27% of total electricity and 50% of low-carbon energy production. In France, nuclear power contributes to 75% of the country's total electricity generation. The present government strategy is to reduce the share to promote renewables. China plans to raise its installed capacity of nuclear power to around 70 gigawatt of electricity (GWe) by 2025. The medium-(2035) and long-term (2050) perspective in China is that of a continued increase in electricity demand, a higher proportion of electricity to reduce fossil energy usage, an accelerated decarbonization of the power supply infrastructure and a rapid development of clean energy. At present, China's energy production mainly relies on coal, which significantly contributes to air pollution and GHG emissions. China has adopted a clean and low-carbon energy strategy aiming to diminish the growth in coal consumption and to reach a peak level in the use of coal as soon as possible, and then proceed to reduce this level by coping with the demand using clean energy. China's key strategic choice in the long-term is to actively develop nuclear power as a pillar of green energy. The French situation features a nearly decarbonated production of electricity mainly based on nuclear and hydro power.

Much of the global energy consumption in the past was dominated by developed countries. However, this is now changing, and the energy landscape is shared by developing countries. The supply of fossil energy can only slowly grow; the energy mix is now challenged by several constraints including resource availability and environmental impact, the need to reduce GHG emissions in relation with climate change, etc. Energy production and utilization is bound to become more efficient, cleaner and low carbon. The global energy infrastructure will be significantly changed in the decades ahead, giving rise to a diversified energy mix in which oil, gas, coal, renewables and nuclear energy will coexist.

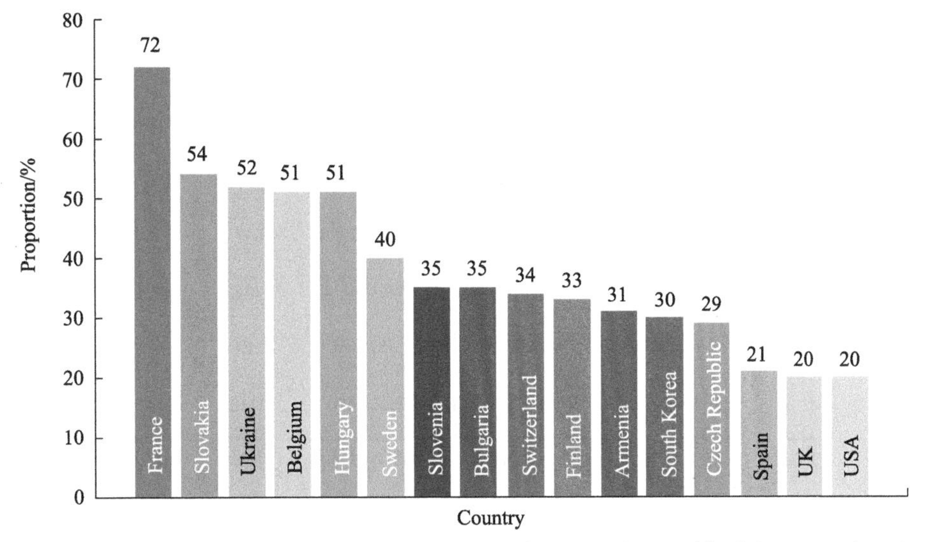


Figure 1 - 1 Countries that nuclear power accounts for more than 20% of domestic electricity supply (December 2016)

In this mix, nuclear power has the advantage of using limited amounts of resources, with a small level of GHG emissions per kWh produced, a relatively low land use and it constitutes an on-demand dispatchable and reliable energy source. The objective of the present report is to develop a comprehensive analysis of its environmental impact, to provide a balanced evaluation of its advantages and weaknesses to consider its sustainability on the long term basis and its capability to respond to the rising demand in electricity.

## 1.2 Decarbonization commitment and $CO_2$ emissions by various energy options

In 1992, the *United Nations Framework Convention on Climate Change* (UNFCCC) put forward a call to stabilize global concentration of greenhouse gases at a level that would prevent a dangerous change in the climate of the planet. Through the *Paris Agreement on Climate Change*, all contracting parties share the objective to control the average global temperature rise and keep it well below 2 ℃ compared to pre-industrial levels, and pursue efforts to limit temperature rise to 1.5 ℃. The United Nations have announced that the Paris Agreement came into effect on November 4, 2016, which laid out arrangements for global actions toward climate change after 2020. France has officially ratified the Agreement on June 15, 2016 as the first industrialized nation and China ratified and acceded to the Paris Agreement on September 3, 2016, as the 23rd contracting party. This Agreement should have a profound impact on the energy mix that will have to shift to essentially low carbon energy sources. Nuclear energy constitutes an important option for achieving low GHG emission levels and complying with climate goals.

Most of the present nuclear energy is based on fission of uranium atoms. The energy released from the fission of 1 kilogram of fissionable material contained in nuclear fuel is equivalent to the energy released from the combustion of 2700 tons of standard coal, indicating that nuclear power is much more efficient and intensive as an energy source than a typical fossil fuel. On a more practical level, this may be illustrated by comparing the amount of fuel that is being used by a typical nuclear power plant to that of a coal fired plant both operating at a power of 1 GWe over a full year. The NPP uses 30 t of fuel while the coal plant requires 4 Mt of coal. The analysis of section 2.3 of this report underlines that nuclear energy emits no particulates, and very limited quantities of air pollutants. However, there are emissions resulting from mining, construction and fuel cycle activities, leading to unavoidable $CO_2$ emissions. In their full life cycles, annual $CO_2$ emissions by NPPs and reprocessing facilities account for less than 1% of those resulting from coal fired power plants, and they are also lower than those associated with production and integration of solar and wind energy supply chains.

## 1.3 Environmental protection as a requirement to make nuclear energy sustainable

In addition to being low carbon and to require a relatively limited amount of land, nuclear energy must be safe and economically competitive. It is however important to examine its environmental impact and jointly consider operation of NPPs as well as that of nuclear fuel cycle facilities. It is also necessary to carefully review the management of the spent fuel that is periodically discharged from the reactor, its storage, possible reprocessing and the final disposal of radwaste. These various facilities generate radioactive gaseous, liquid and solid waste. Gaseous and liquid effluents are processed and stored until they have reached regulatory levels which allow their release into the environment. Solid waste is processed and provisionally stored to reduce their volume and activity, to comply with the requirements of the waste minimization standards; it is temporarily stored or directly sent to final disposal.

According to the safety analysis and the environmental impact assessments, the authorized releases from facilities result in doses (radiological and chemotoxic) to people at levels that are lower than what is specified in the regulatory requirements. The national regulations are at least compliant with IAEA requirements but very often more stringent. The target is to remain far below the individual effective dose limit of 1 mSv per year to the representative person, as recommended by the International Commission on Radiological Protection (ICRP). The authorized limit and practical objectives for the release of effluents are becoming lower and lower and this has led to continuous improvement of the treatment processes.

In the first report in 2017, the three partnering Academies have analyzed some issues and challenges raised by nuclear energy, with regard to safety, management of radwaste, development and deployment of advanced nuclear energy systems, economics, public acceptance, etc. The environmental impacts of nuclear energy were left aside in comparison with the topic of nuclear energy safety. However, the public is progressively becoming more sensitive to the global and local impacts on the environment from industrial activities and in particular from those required to produce massive amounts of electrical energy. Global impacts take more importance and will drive the future choices of energy mixes. Energy and ecologic transitions become inseparable.

Local immediate or deferred environmental impacts are major components of related social issues that may influence the acceptance or rejection of nuclear energy. It is important to provide a comprehensive evaluation of how nuclear energy impacts the environment, given the considerable attention on how these issues receive, and thereby offer a complete and balanced account of measures taken to limit such negative impacts. This will hopefully be a useful addition to the first report of our Academies and will allow a better assessment of this topic. The sustainability of nuclear energy also depends on the confidence that the nuclear countries can have towards the newcomers in their capacity to adhere to the principles of environmental impact control.

A key objective of the second report is to assess nuclear energy potential as ways to produce clean energy. To consider this question, the Academies decided to look at different indicators used to measure the impacts at global and local scales in constructing, operating and dismantling NPPs and fuel cycle facilities considering all situations (normal and accidental). All impacts are investigated along the processes implemented at each step of the front-end (from mining to fabrication of fuel assemblies) and back-end (from waste management to dismantling of NPPs and facilities) of the nuclear fuel cycle. As radioactivity is always present in nuclear energy production, attention will be paid firstly to the risk of exposure to ionizing radiation of living beings over extended periods of time.

## 1.4 Report contents and organization

This report comprises an executive summary including recommendations and six chapters.

The next chapter discusses environmental consequences during normal operations of NPPs and fuel cycle facilities. It includes a comparison of various electricity production systems in terms of greenhouse gas and atmospheric pollutant emissions and then discusses issues related to radioactivity associated with normal operation, water consumption, land use and material requirements.

Chapter 3 considers spent fuel and radwaste management. It introduces the principles, strategy and framework aimed at preventing environmental impacts. Basically, this management distinguishes various classes of radwaste, the processing and discharge and disposal of radioactive waste, and the different impacts related to the open and closed nuclear fuel cycle. The environmental protection measures that are taken at each step of radwaste management are discussed.

Chapter 4 reviews severe nuclear accidents (TMI, Chernobyl and Fukushima) to underline lessons learned from these events. It describes upgrades that have been introduced in existing NPPs and improvements that are included in the new Gen-Ⅲ designs in order to limit to the nuclear site boundary, the environmental impacts in cases of accident.

Chapter 5 describes nuclear safety in relation to the environment. It discusses the objectives of nuclear safety that are to restrict the likelihood of a nuclear accident and the prevention and mitigation of the consequences. It considers the problem of NPPs sitings, the role of safety authorities, the responsibility of nuclear plant operators and that of the government.

Chapter 6 summarizes the main findings of the study. References and a glossary can be found at the end of the report.

# Chapter 2　Environmental impacts during normal operations of nuclear power plants and nuclear fuel cycle facilities

---

Recommendations

The exposures resulting from nuclear activities must always be compared to natural exposures and to exposures resulting from other electricity producing technologies.

Although the large majority of epidemiological studies around the world converge to demonstrate that the long-term effects of low and very low dose/dose rate exposures are not harmful, it is still recommended that background epidemiological studies be implemented before the operation of a new nuclear facility, which can provide valuable information for the comparative analysis of post-accident epidemiology studies, analysis of radiation risks, and response to public concerns.

Fast neutron reactors and multi-recycling have the potential to drastically reduce the environmental footprint of nuclear energy, by reducing uranium mining activities, and the quantity and toxicity of nuclear waste. Although this technology is not required in the immediate future, preparation of the commercial development of fast neutron reactors in the coming decades must be pursued.

---

## Introduction

This chapter deals with the environmental footprint of nuclear energy, compares the impacts of this system of production of electricity with other electricity generation systems and discusses some trends regarding the reduction of impacts resulting from the introduction of new nuclear technologies. It begins with some general considerations about the environmental impacts expected from nuclear energy in the framework of human activities.

## 2.1 How to measure impacts of nuclear energy on the environment

Environmental impacts are temporary or permanent modifications of given parts of our natural environment, including air, water, land, flora, wildlife, etc., with ourselves potentially as an ultimate target. They can, for example, be caused by releases of gas, liquids or solids from human activities.

### 2.1.1 Main impacts of nuclear energy to the environment

The main environmental impacts are related to climate change ($CO_2$ and other GHG emissions), air and water pollutions (different releases), water withdrawal and water consumption, creation of man-made land or loss of heritage, degradation of natural land, soil erosion, consumption of raw material, production, processing and disposal of waste …

Climate change is considered as having the most severe impact on global environment; furthermore, this is accompanied by marine and terrestrial ecosystems degradation, and the loss of biodiversity. These degradations come from acidification and eutrophication linked to emissions of gaseous sulfur and nitrogen oxides together with $CO_2$. Contributions of nuclear energy to these emissions are very low. Regarding other impacts, such as land occupation, the water cycle, the contributions are variable according to the fuel cycle options and will be considered further.

Regarding nuclear energy, the releases to the environment are radioactive or suspected to be so. They can generally lead to exposure of human and other living beings to radiation. The potential impacts of these cumulative exposures have to be assessed with respect to human health and biodiversity. To perform these assessments, exposures must always be compared with those of natural sources of radiation, or to those used for medical diagnostic.

(1) Exposures to natural background radiation: the annual average effective dose to the public from natural background radiation for example in France, the USA and China is 2.9 mSv (Ref. [2]) 3.1 mSv (Ref. [3]) and 3.1 mSv (Ref. [4]) respectively; in some high background radiation areas, the dose level can be much higher such as in Kerala (India) where it is more than 10 mSv. A New York-Paris round trip flight would expose a person to about 0.05 mSv (Ref. [5]).

(2) Exposures to other natural sources of radiation caused by human activities like those associated with rare earth extraction and processing or those induced by coal-fired power generation. Massive utilization of slag as building material leads to a significant increase of exposure to indoor radon in China (Ref. [4]).

(3) Exposure to medical diagnostic causes about 1 mSv per year on average (rounded value from Ref. [5], page 54).

### 2.1.2 Releases to the environment

The releases to the environment fall into two large categories:

(1) Immediate releases of radioactive substances from NPPs and facilities leading to their dispersion, dilution, deposition on soils, lixiviation from soils, migration in soils, the driving forces being wind and rain.

(2) Long term releases of radioactive substances from waste packages (leaching or dissolution of conditioning materials), leading to their migration as true species or colloids, the driving force being natural geosphere gradients (hydraulic, thermal, chemical).

### 2. 1. 3 Assessments

There are two ways of assessing the impacts on the environment associated with large energy production systems, depending on the time and scale considered.

When one considers large potential impacts, like those affecting the atmosphere and impacts extending over long periods of time, it is appropriate to use life cycle analysis (from cradle to grave) starting with the construction and ending with the dismantling of the nuclear facility (over about a century) to evaluate global impacts. LCA accounts for the impacts already recorded and for the expected ones. Results of LCA support for instance the figures given in Section 2. 2 and 2. 4 of this chapter.

When the impact is limited to and around the sites of the facilities and when the time refers to "daily life" (local impacts), both immediate and long-term deferred impacts are important; but only the former can be measured, while the latter need to be obtained from simulations. This approach is used for instance in the upcoming Section 2. 3.

In some cases, natural analogues provide experimental data on long-term impacts, for example, the limited migration of radionuclides during billions of years in the Oklo uranium deposit (Gabon- (Ref. [6]) ) or reduced alteration of the surface of glasses in the Mediterranean Sea during thousands of years, which have protected these glasses from dissolution.

*Assessments consider all the radionuclides and other elements both natural and man-made present in the releases.*

*Nuclear instrumentation can detect and characterize very low levels of radioactivity*, which is less the case for chemotoxicity. Individuals or associations can easily find basic nuclear instrumentation at low cost and do their own *in-situ* or remote measurements after an appropriate but short training to assure a correct level of quality. Thus, radioactivity measurements are more and more as a domain where independent studies can be done, assuring independence and cross checking of the levels and nature of radioactivity in the environment.

The tools for measuring trace amounts of chemotoxic substances in the environment are more complex than those for measuring radioactivity and the *in-situ* acquisition of chemotoxic data is therefore difficult.

In France and other European and probably also non-European countries where nuclear facilities exist, *a detailed analysis of all the materials, sources and wastes, existing in the nuclear, industrial or medical facilities of the country, is done periodically and a programme for managing them on the short-and long-term is implemented*. Accordingly, assessments of real or potential radioactive releases can be done.

### 2.1.4 Methodology

In each domain of interest where impacts are expected, some parameters that measure the various potential or real detrimental effects can be selected for comparing the environmental impacts of different energy systems.

Then LCA can be implemented according to energy production scenarios and the characteristics of the respective energy systems.

The estimation of local impacts is possible when taking into account the characteristics of the facilities and targets considered and the way of living of local populations.

Radiological and chemical impacts on humans or biotopes may be estimated using simulations. All the codes follow more or less the same steps and rely on large databases. However, they require adequate validations, and some results can be subjected to debate. Results from long-term (thousands of years) impact simulations can be even more debatable.

## 2.2 Effluents, radiological impacts of nuclear energy and solutions

### 2.2.1 Effluents and radiological impacts of NPPs

The radiological impacts of NPPs in France during normal operation are shown in

Table 2 - 1 (Ref. [7]). The estimated dose levels are well below, by more than three orders of magnitude, the natural radioactivity dose levels mentioned previously.

**Table 2 - 1 Radiological impacts of NPPs since the year of 2011 calculated on the basis of the actual discharges from the installations and for the most exposed reference groups**

| EDF NPP | Distance to site/km | Estimation of received doses/(mSv/a) | | | | | |
|---|---|---|---|---|---|---|---|
| | | 2011 | 2012 | 2013 | 2014 | 2015 | 2016 |
| EDF/Belleville-sur-Loire | 1.8 | $8\times10^{-4}$ | $8\times10^{-4}$ | $7\times10^{-4}$ | $4\times10^{-4}$ | $5\times10^{-4}$ | $4\times10^{-4}$ |
| EDF/Blayais | 2.5 | $6\times10^{-4}$ | $2\times10^{-4}$ | $2\times10^{-4}$ | $6\times10^{-4}$ | $5\times10^{-4}$ | $5\times10^{-4}$ |
| EDF/Bugey | 1.8 | $8\times10^{-4}$ | $5\times10^{-4}$ | $6\times10^{-4}$ | $2\times10^{-4}$ | $2\times10^{-4}$ | $9\times10^{-5}$ |
| EDF/Cattenom | 4.8 | $3\times10^{-4}$ | $3\times10^{-3}$ | $5\times10^{-3}$ | $8\times10^{-3}$ | $7\times10^{-3}$ | $9\times10^{-3}$ |
| EDF/Chinon | 1.6 | $5\times10^{-4}$ | $5\times10^{-4}$ | $3\times10^{-4}$ | $2\times10^{-4}$ | $2\times10^{-4}$ | $2\times10^{-4}$ |
| EDF/Chooz | 1.5 | $1\times10^{-3}$ | $9\times10^{-4}$ | $2\times10^{-3}$ | $7\times10^{-4}$ | $6\times10^{-4}$ | $6\times10^{-4}$ |
| EDF/Civaux | 1.9 | $7\times10^{-4}$ | $9\times10^{-4}$ | $2\times10^{-3}$ | $8\times10^{-4}$ | $9\times10^{-4}$ | $2\times10^{-3}$ |
| EDF/Cruas | 2.4 | $5\times10^{-4}$ | $4\times10^{-4}$ | $4\times10^{-4}$ | $2\times10^{-4}$ | $2\times10^{-4}$ | $2\times10^{-4}$ |
| EDF/Dampierre-en-Burly | 1.6 | $2\times10^{-3}$ | $1\times10^{-3}$ | $9\times10^{-4}$ | $4\times10^{-4}$ | $5\times10^{-4}$ | $5\times10^{-4}$ |
| EDF/Fessenheim | 3.5 | $8\times10^{-5}$ | $1\times10^{-4}$ | $1\times10^{-4}$ | $4\times10^{-5}$ | $4\times10^{-5}$ | $3\times10^{-5}$ |
| EDF/Flamanville | 0.8 | $2\times10^{-3}$ | $6\times10^{-4}$ | $7\times10^{-4}$ | $5\times10^{-4}$ | $2\times10^{-4}$ | $2\times10^{-4}$ |
| EDF/Golfech | 1 | $8\times10^{-4}$ | $7\times10^{-4}$ | $6\times10^{-4}$ | $2\times10^{-4}$ | $3\times10^{-4}$ | $3\times10^{-4}$ |
| EDF/Gravelines | 1.8 | $2\times10^{-3}$ | $4\times10^{-4}$ | $6\times10^{-4}$ | $8\times10^{-4}$ | $4\times10^{-4}$ | $4\times10^{-4}$ |
| EDF/Nogent-sur-Seine | 2.3 | $8\times10^{-4}$ | $6\times10^{-4}$ | $1\times10^{-4}$ | $5\times10^{-4}$ | $4\times10^{-4}$ | $7\times10^{-4}$ |
| EDF/Paluel | 1.4 | $8\times10^{-4}$ | $5\times10^{-4}$ | $9\times10^{-4}$ | $9\times10^{-4}$ | $4\times10^{-4}$ | $3\times10^{-4}$ |
| EDF/Penly | 2.8 | $1\times10^{-3}$ | $6\times10^{-4}$ | $7\times10^{-4}$ | $4\times10^{-4}$ | $4\times10^{-4}$ | $4\times10^{-4}$ |
| EDF/Saint-Alban | 2.3 | $4\times10^{-4}$ | $4\times10^{-4}$ | $4\times10^{-4}$ | $2\times10^{-4}$ | $2\times10^{-4}$ | $3\times10^{-4}$ |
| EDF/Saint-Laurent-des-Eaux | 2.3 | $3\times10^{-4}$ | $2\times10^{-4}$ | $2\times10^{-4}$ | $2\times10^{-4}$ | $1\times10^{-4}$ | $1\times10^{-4}$ |
| EDF/Tricastin | 1.3 | $7\times10^{-4}$ | $7\times10^{-4}$ | $5\times10^{-4}$ | $2\times10^{-4}$ | $2\times10^{-4}$ | $2\times10^{-4}$ |

Monitoring results of gaseous and liquid effluents during the operation of six pressurized water reactors (PWRs) NPPs and one heavy water reactor (HWR) NPP in China were analyzed, and Figure 2 - 1 shows the average emission of various type of effluents during years of 2011 to 2013 of these seven NPPs, of which the maximum emissions are effectively regulated and controlled; in all cases, it is well below the regulatory limits and the natural exposure. Normalized collective dose to the public from effluents of NPPs in China during the years of 2011 to 2013 was estimated as $6.4\times10^{-2}$ man · Sv/GWa.

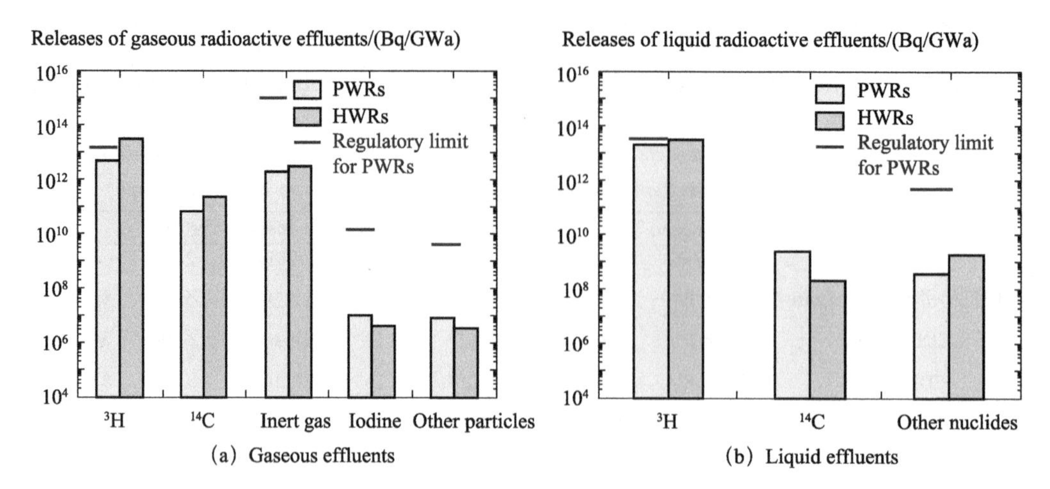


Figure 2 - 1 The average emission of effluents of NPPs in China (2011-2013) (Ref. [8])

It is noted that, tritium (T) is one of the radionuclides released by NPPs to the environment. As an isotope of hydrogen, its behavior in the environment is mainly linked to the water cycle (HTO or tritiated water) but also to photosynthesis (incorporation of $T_2$ or HT molecules in plants) and to the metabolism of organic tritiated molecules in living organisms (organically bound tritium or OBT). The World Health Organization recommends a guidance level of 10000 Bq/L for tritium in drinking water for permanent consumption (Ref. [9]). Reports from the French Institute of Radiation and Nuclear Safety (IRSN) indicate that there is no evidence of tritium bio-accumulation in vegetal components after decades of study in France (IRSN-Ref. [10]). For terrestrial animal products, the conclusion remains the same, but based on limited data (Ref. [11]) as most studies are focused on the physiological models describing the behavior of tritium in the animals in view of estimating the concentration in the animal products (milk, meats, etc.), however, the recent estimated transfer factors (Ref. [11]) are always less than 1, which confirms an absence of accumulation in food originating from animals.

There is also no evidence of bio-accumulation of tritium in seawater species (IRSN-Ref. [12]).

Over the years, and as illustrated in Figure 2 - 2, the liquid and gaseous radioactive releases have been drastically reduced, both in China and France.

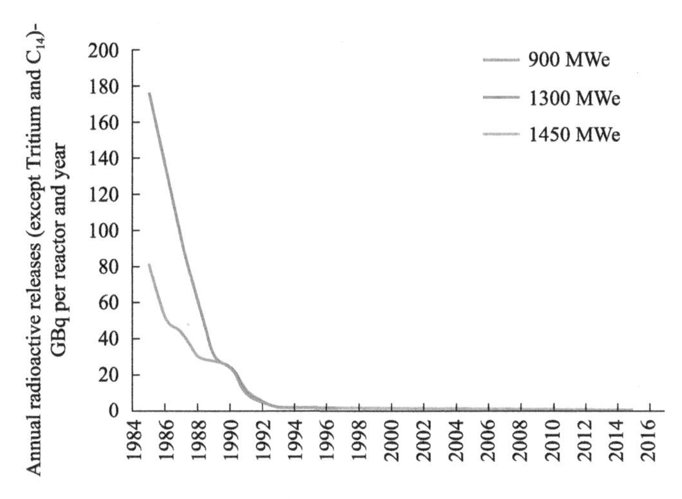


Figure 2 - 2 Liquid releases of French NPP (1984-2015)

## 2. 2. 2 Effluents and radiological impacts of nuclear fuel cycle

Nuclear fuel cycle includes the production, fabrication, storage and post-processing activities of nuclear fuel. The estimated radiological impacts of nuclear fuel cycle in France during normal operations are shown in the Table 2 - 2 (Ref. [7]). The estimated received doses are again quite low, from two to four orders of magnitude below the annual dose from natural radioactivity.

**Table 2 - 2 Radiological impacts of nuclear fuel cycle plants since the year of 2011 calculated on the basis of actual discharges from the installations and for the most exposed reference groups**

| Nuclear fuel cycle plants | Distance to site/km | Estimation of received doses/(mSv/a) | | | | |
|---|---|---|---|---|---|---|
| | | 2011 | 2012 | 2013 | 2014 | 2015 |
| Andra/CSA | 2. 1 | $3\times10^{-4}$ | $1\times10^{-5}$ | $1\times10^{-6}$ | $2\times10^{-6}$ | $2\times10^{-6}$ |
| Andra's Manche repository | 2. 5 | $6\times10^{-4}$ | $4\times10^{-4}$ | $4\times10^{-4}$ | $3\times10^{-4}$ | $2\times10^{-4}$ |
| Areva NP in Romans F | 0. 2 | $6\times10^{-4}$ | $6\times10^{-4}$ | $5\times10^{-4}$ | $3\times10^{-4}$ | $3\times10^{-4}$ |
| Areva/La Hague | 2. 8 | $9\times10^{-3}$ | $9\times10^{-3}$ | $2\times10^{-2}$ | $2\times10^{-2}$ | $2\times10^{-2}$ |
| Areva/Tricastin | 1. 2 | NA | $3\times10^{-4}$ | $3\times10^{-4}$ | $3\times10^{-4}$ | $3\times10^{-4}$ |

The effluents of nuclear fuel cycle in China are well controlled and documented. Figure

2 - 3 and Figure 2 - 4 show effluent emissions and their radiological impacts to the public of the nuclear fuel cycle in China from the year of 2011 to 2013, respectively.

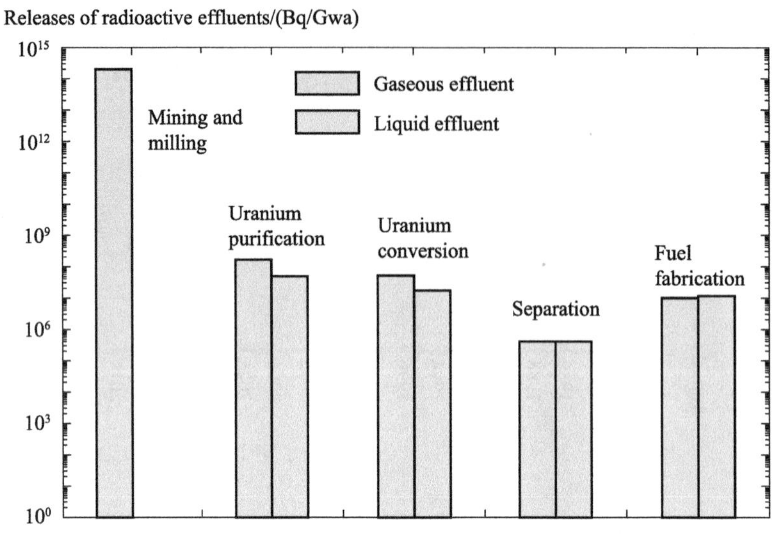


Figure 2 - 3 Average emission of effluents of nuclear fuel cycle in China (2011-2013) (Ref. [8])

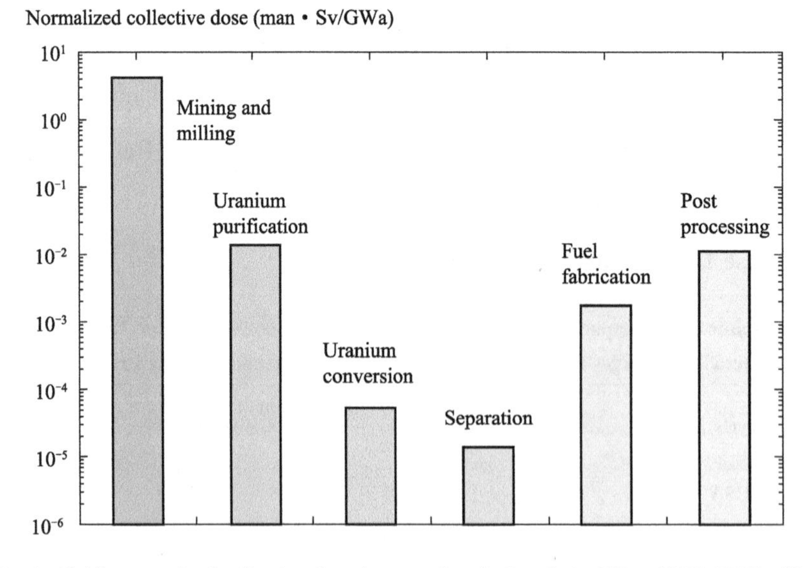


Figure 2 - 4 Public normalized collective dose from nuclear fuel cycle in China (2011-2013) (Ref. [8])

(Mining and milling data refer to $^{222}Rn$ emission; others refer to total uranium)

In addition, public normalized collective dose from LCA of nuclear power generation in China from the year of 2011 to 2013 was estimated at 4. 6 man • Sv/GWa (80 km distance to the site, detailed in Appendix 2-1, as well as radiological impacts resulting from other power generation technologies), of which 86% was contributed by uranium mining and milling. As the *in-situ* leaching uranium mining technologies get expanded in the near future in China, the dose to the public from nuclear power chain will be further reduced.

### 2.2.3 Radiation monitoring and surveillance

Monitoring of the radioactivity in the environment is the main concern of all operators and safety and environmental authorities in all nuclear countries. Here, we focus on the French and Chinese systems.

In France, monitoring of radioactivity is implemented through three tele-surveillance networks:

(1) The system of drills (water monitoring) and beacons (air monitoring) installed and operated by the owner/operator of the nuclear site in the vicinity (0-10 km in France) of the site.

(2) The network for monitoring radioactivity in the air, the purpose of which is to immediately detect any unusual increase of radioactivity in the air: in France, Teleray includes 400 beacons spread over the national territory with special focus on large cities situated less than 30 km of a nuclear site.

(3) The network for continuously monitoring radioactivity of water in the seven main French rivers upstream of their estuaries or of their point of passage into a neighboring country: limits of detection are very low, in the range of 0.5 to 1 Bq/L for cesium-137, iodine-131 and cobalt-60.

In addition, networks of surveillance by sampling are set up to evaluate the impact on ambient air of all human activities using radionuclides: OPERA-AIR operated by the French IRSN, with 40 stations including 32 in the vicinity of nuclear sites, sampling of water, mud and sediments, and milk.

The Chinese surveillance system consists of a multi-level environmental radiation monitoring network, so as to keep detecting environmental radiation levels during the operation of nuclear facilities.

(1) Immediate site vicinity (generally within a 5 km radius from an NPP site): the fixed automatic monitoring stations (autonomous monitoring), which are set up, run and managed by the operator.

(2) Long range distance to sites: (generally 20 km radius away from the site): the fixed automatic monitoring stations (supervising monitoring), which are set up by the operator, but run and managed by the provincial environmental protection administration departments. The autonomous monitoring is combined with the supervising monitoring to measure the environmental $\gamma$ radiation level and sampling the air in the respective areas.

(3) Intermediate distance to a site (generally 10 km radius away from the site, including the inner area): Environmental material samples, such as surface water, underground water, receiving water, soil, and bottom mud, are monitored and analyzed by theoperator.

In major cities and regions of the nation, the environmental medium, such as air, water and soil, should be monitored, sampled and analyzed by the national radioactive environmental monitoring stations.

### 2.2.4 Biological effect of ionizing radiation

Deterministic effects are induced by ionizing radiation higher than an established threshold, while stochastic effects, generally cancer or heritable diseases, might be induced at low doses and low dose rates.

In China, since 1972, the possible health effects on large populations induced by ionizing radiation exposure at low dose and low dose rate, has been investigated in high background radiation area (HBRA) in Yangjiang, Guangdong province, and in normal background radiation in control area (CA), of which the annual average natural radiation dose to inhabitants was estimated at 6.4 mSv and 2.4 mSv respectively (revised to 5.9 mSv and 2.0 mSv, respectively, in 2000).

No harmful impacts were found by natural radiation in HBRA, based on the investigation on cancer mortality from 1008769 person-year in HBRA and 995070 person-year in CA, and on heritable diseases, congenital malformations, chromosome aberrations and immune function of peripheral blood lymphocytes from 13425 persons in HBRA and 13087 persons in CA.

In European and American countries, since the 1950s, epidemiological investigations have been performed near nuclear facilities. However, the results showed that no significant difference could be found in terms of cancer mortality and childhood leukemia incidence near nuclear facilities as compared to control areas. This is mainly due to very low doses to the public induced by the radioactive emissions during normal operation of nuclear facilities. The additional dose received by the critical group is as low as about 10 μSv in one year, about 1% of the natural background radiation level (the order of 1 mSv/a) excluding radon exposure.

There exist large uncertainties to low doses of ionizing radiation in the mechanisms of cancer induction and the biological effect, and the dose assessments to the public near nuclear facilities, thus it is difficult to reach quantitative conclusions on radiation risks by epidemiological investigations, making it useless to implement large scale conventional studies. New approaches based on biochemical signature of radiation effects are expected but are not yet operational.

However, implementing background epidemiological studies before the operation of a new nuclear facility could provide very valuable information that would allow comparative analysis to be carried out after an accident. This will help assessing radiation risks, and responding to the concerns about the consequences of higher exposure doses to the public around the nuclear facility resulting from the eventual release of radioactive substances during and after an accident.

More results on epidemiological investigations are given in Appendix 2 - 2.

### 2.2.5 Transportation of radioactive materials

Around 900000 radioactive packages are transported in France every year for the needs of the industry, the medical sector or research; the bulk of it handles very low sources and wastes. Only 15% are related to fuel and low, intermediate or high-level radioactive waste. For the whole world, the number of nuclear packages reaches 10 million, which represents only 2% of the total of all hazardous material packages transported.

With respect to humans and the environment, the main risks are irradiation and contamination. In France, one to two transportation accidents occur per year inducing radioactive releases to the environment. They all have had limited impacts; in the most serious cases, weak contamination has been detected and treated by local decontamination operations.

Railway transportation is the priority means with a very high level of safety for heavy or bulky packages.

Maritime transportation is used for around 4% of the total transportation of nuclear materials, and mainly for fresh or spent fuel and high-level radioactive waste. Ships are specifically designed according to the requirements of the International Maritime Organization.

Road transportation is the most flexible means to transport radioactive materials. It is submitted to special rules to avoid crowded periods and housing areas.

Air transportation is used only for small and urgent packages, such as radiopharmaceuticals, and over long distances.

Suitable options should be chosen based on the characteristics of radioactive materials and the requirements of transportation.

### 2.2.6 Participation of stakeholders

In France, local commissions for information of stakeholders (CLI) are set up for the most hazardous facilities classified as important for environmental protection (ICPE).

53 CLIs exist in France, including 38 around nuclear sites. They bring together around 3000 members: local politicians, trade unions, representatives of associations, experts and qualified persons. They have the general mission of informing the public about the safety of the facilities classified as above and their impact on persons and the environment. In the nuclear domain, the *Transparency and Security Act* (June 13,

2006) gave them a legal basis (Art. 125-7 of the *Environmental Code*).

A National Association of the CLIs (ANCCLI) gathers the experience and wishes of 37 CLIs and brings their collective insights to the attention of national and international authorities.

China has established a public communication mechanism that combines supervision by the central government, guidance by local governments, implementation by enterprises, and participation of the public to promote popularization of science, public participation, information publicity, public opinion response and integrative development.

The *Nuclear Safety Law* is the legal basis and guarantee of the public's rights to know about, participate to and supervise major nuclear energy projects. The development of major new nuclear projects is incorporated into the review system of local people's congresses, and public communication on major nuclear related projects is included into the local social management system. Enterprises are required to develop public communication strategies and medium-and long-term plans, into their operational management. In addition, peer review is carried out by professional and authorized third parties such as social organizations, universities and think tanks, for example the China Nuclear Energy Association, the Chinese Nuclear Society, the China Association for Science and Technology and the China Environmental Protection Association.

## 2.3 Environmental impacts of nuclear energy compared to other sources of electricity

Power generation has environmental impacts which depend on technologies.

Within the framework of the transition to a decarbonized economy the objective of this section is to look at the non-radiological environmental impacts of various power generation technologies, for example the GHG emissions, land occupation, material consumption for construction, water consumption and decommissioning waste (the radiological impacts of nuclear energy are detailed in section 2.2). Most of these figures originate from life cycle analysis (LCA). Appendix 2-1 summarizes the recent analysis results on the life cycle GHG emissions and radiological impacts of various power generation technologies in China.

Figure 2-5 below (Ref. [13]) shows that $CO_2$ emissions from fossil fuel fired electrical power plants are one to two orders of magnitude higher per MWh produced than those from nuclear, wind, solar and hydro.

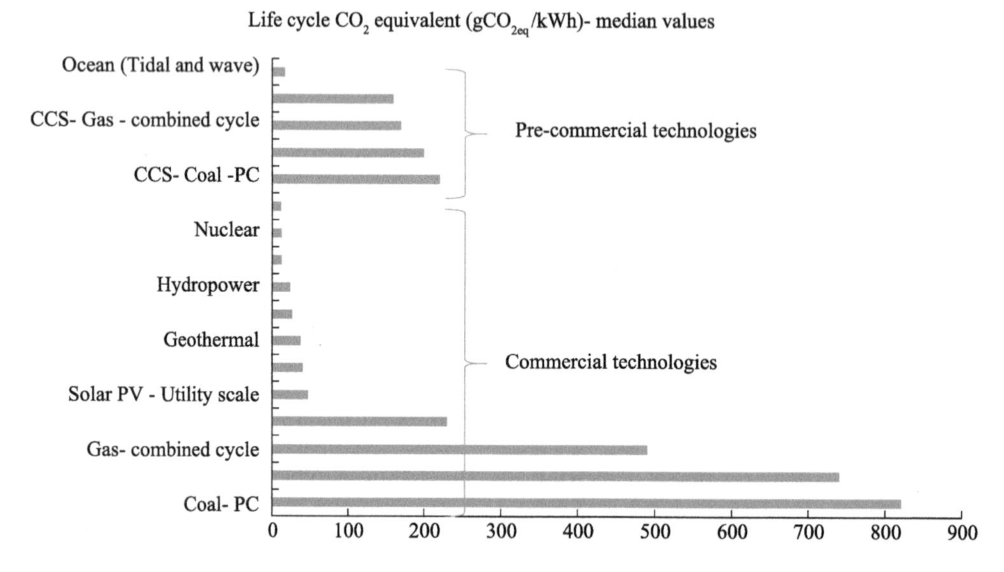


Figure 2-5 Life cycle $CO_2$ equivalent from selected electricity supply technologies. Arranged by decreasing median ($gCO_{2eq}$/kWh) values

The contribution of nuclear energy to the emission of $SO_x$ and $NO_x$ gases (around 20 kg/MWh), is two and one order of magnitude less than those from fossil fuels and photovoltaic electricity respectively and consequently has a low impact on acidification of soils and eutrophication of water. $SO_x$ and $NO_x$ emissions from hydro- and wind-power are less than 10 kg/MWh.

Wind, solar, nuclear, and hydro-power have their own environmental impacts. We shall look at the following impacts:

(1) Land occupation;

(2) Material usage for construction;

(3) Water consumption;

(4) Deconstruction waste.

Wind and solar, as intermittent sources of electricity, should also be assigned $CO_2$ emissions from fossil fuel facilities which have to be operated as back-ups when they are not available.

### 2.3.1 Land occupation

Many studies and research projects have addressed the land-related impacts on energy systems which, increasingly, are focusing on "renewables". Drawing from this body of knowledge, Table 2 - 3 (Ref. [14] and [15]) provides a brief synthesis of the land footprint relating to these systems on a MW basis (LCA-life cycle analysis).

Table 2 - 3 Land use intensity per MW of installed capacity

| Energy technology | $m^2/MW$ | System boundary<br>Energy resource extraction area plus power plant site |
|---|---|---|
| Hydropower reservoir | 20000 - 10000000 | Site of reservoir and generators |
| Solar PV | 10000 - 60000 | Site of PV system, which includes the area for solar energy collection. PV systems on pre-existing structures have essentially no net increase in land use |
| Solar thermal | 12000 - 50000 | Site of concentrating solar thermal system, which includes the area for solar energy collection |
| Wind | 2600 - 1000000 | Low-end value is for the site only, which includes the physical footprint of the turbines and access roads. The high-end value includes the land area between turbines, which is typically available for farming or ranching |
| Nuclear | 6700 - 13800 | Low estimate is site only. High estimate includes transmission lines, water supply, and rail lines, but does not include land used to mine, process, or dispose of waste |

The figures proposed for hydro seem to be high; however, power generation, in many cases, is but one of the various purposes of the dam (water storage for irrigation, domestic and industrial uses, shipping, flood protection). The land use is due to the extent of the reservoir where one is needed. Regarding energy supply, the purpose of the reservoir is not only the delivery of power but also flexible storage of electricity, thus creating added value.

The direct land use from NPPs is very low, thus nuclear power is a favorable option with regard to land use, and consequently to preserve biodiversity which is undermined by land occupation and artificialization.

It is noted that the global perception of nuclear energy is influenced by the footprints generated in case of an accident, as occurred in Chernobyl and Fukushima (see Chapter 4). Public opinion is legitimately concerned with restriction of use of radioactive areas following severe nuclear accidents. One of the consequences has been to devise technologies that would confine the impact of an accident to the premises of the nuclear site and thus to avoid any evacuation (see Chapter 4). Nuclear accidents have local negative externalities that need careful follow-up actions. It is worth noting that GHG emissions concern the whole planet and constitute a negative externality that cannot be localized.

### 2. 3. 2 Materials used for construction

The civil works forNPPs require more concrete and steel than a coal fired power plant or a CCGT power plant of the same installed power: around 600 tons per MW compared with around 10 tons per MW respectively. This is due to the para-seismic design of safety-classified buildings, the containment system, the protection shell against airplane crash and the complex concrete raft where a containment tank (core catcher) is designed to collect and cool the core in the case of accidental melting. Some of these features are specific for NPPs of the third generation. However, the quantity of concrete and steel becomes less important when relating it to the amount of kWh produced by a single site during the life span of 60 years.

Onshore wind farms require somewhat more concrete per MWh than NPPs because of their relatively lower load factor. Offshore wind farms laid on the subsea ground require much more cement, aggregate and rockfill for the construction of the foundation on the subsea floor if a gravity basement is selected. The present experience of floating offshore wind farm is too brief to produce a sound figure for the required anchorage foundations. However, the volume of aggregate and rock fill will be far lower than for masts laid on the subsea ground.

The solar PV farms as well as concentrated solar energy farms require steel and concrete. In both cases, there are slabs of reinforced concrete, and steel supports. The low concentration of power associated with a low load factor give rise to a relatively high demand of material per MWh delivered.

There is a large requirement for copper and aluminum in the connection system within the site and also for connecting the site to the grid. The need for copper is high especially for offshore wind farms which have to be connected between themselves and to the shore.

The estimates for hydro are not very relevant for the reasons mentioned in § 2.3.1, and because the need for aggregate and cement depends on the availability of rock and aggregate close to the dam site. Many large dams are embankment dams, because they are the cheapest solution. The volume of concrete of this type of design remains, however, significant and depends on the size of the maximum flood and the installed power.

According to several reports ((Ref. [15] and [16]), this factor leads to the following ratio of material used per TWh of electricity produced on an LCA basis (Figure 2-6).

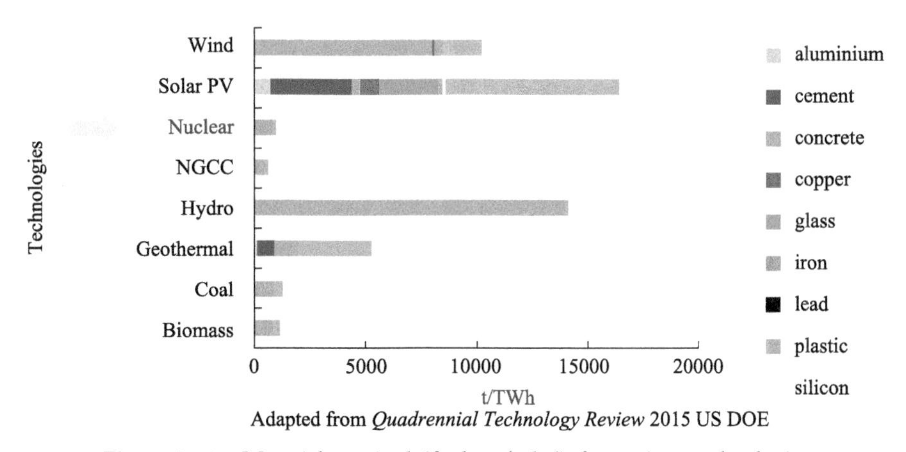


Figure 2-6 Material required (fuel excluded) for various technologies

### 2.3.3 Water withdrawal and consumption

The NPPs require large quantities of water to condense the steam driving the main turbine. Because of the Carnot rule which applies to all thermal plants, roughly one third of the thermal energy of the reactor is converted to electricity, and two thirds is dumped to the environment. This paragraph provides a brief review of the environmental consequences of this issue, which depend on selected technologies.

(1) Many NPPs are built close to the seashore, and are cooled with sea-water. The temperature of the cooling sea water increases by ~7 ℃ in the condenser before being returned to the sea: surface seawater heating after first dilution does not exceed 1 ℃ in an area that can vary from 1 to 20 $km^2$ (Ref. [17]). The cooling water flow has to be sufficient to ensure a temperature increase that respects the needs of aquatic life in the vicinity.

(2) When NPPs are built inland near large rivers, two technical options may be used (Figure 2-7).

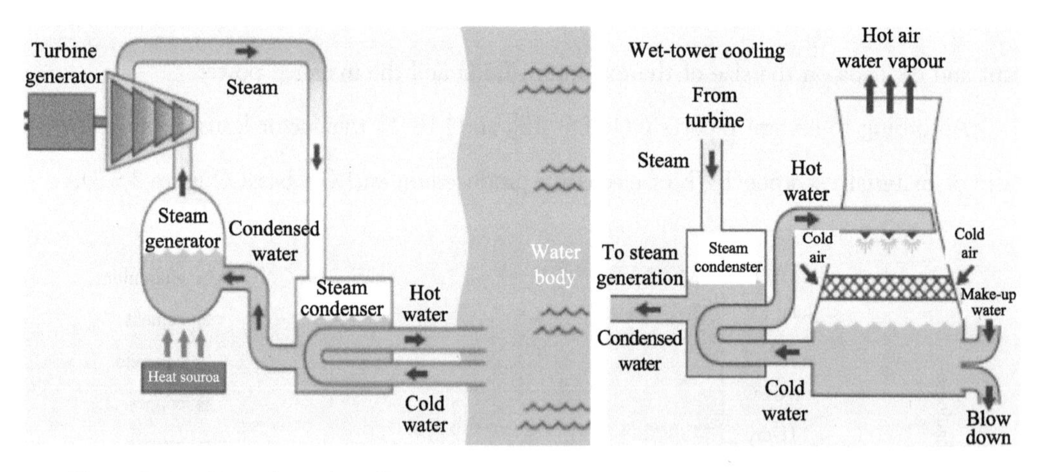


Figure 2-7 Once through cooling systems—Wet tower, cooling system-Courtesy of SFEN

● Once through cooling system: the cooling water withdrawn from the river is returned after having cooled the condenser. A large water flow is necessary to ensure a limited temperature increase. But actual water consumption is limited to extra evaporation of water returned to the river with an increased temperature. Large amounts of cooling water are needed, but water consumption is significantly smaller than withdrawal.

● Wet tower cooling system: part of the water having cooled the condenser is steamed to the atmosphere, and is diverted from the other water demands.

As the latent heat of water is much higher than its sensible heat (five times more energy is needed to vaporize one liter of water, than to heat it from 0 ℃ to 100 ℃), wet tower cooling withdraws less water than once-through cooling, but more water is lost.

The following data issued from a study on "Life cycle water use for electricity generation: a review and harmonization of literature estimates" (Ref. [18]) in the US show that water withdrawals and water consumption vary widely according to the power technology selected and also according to the water scheme retained for each technology.

The range of water consumption[①] for each technology is summarized in the following sketch (Figure 2 - 8) where CSP means concentrating solar power and PV means photovoltaic.

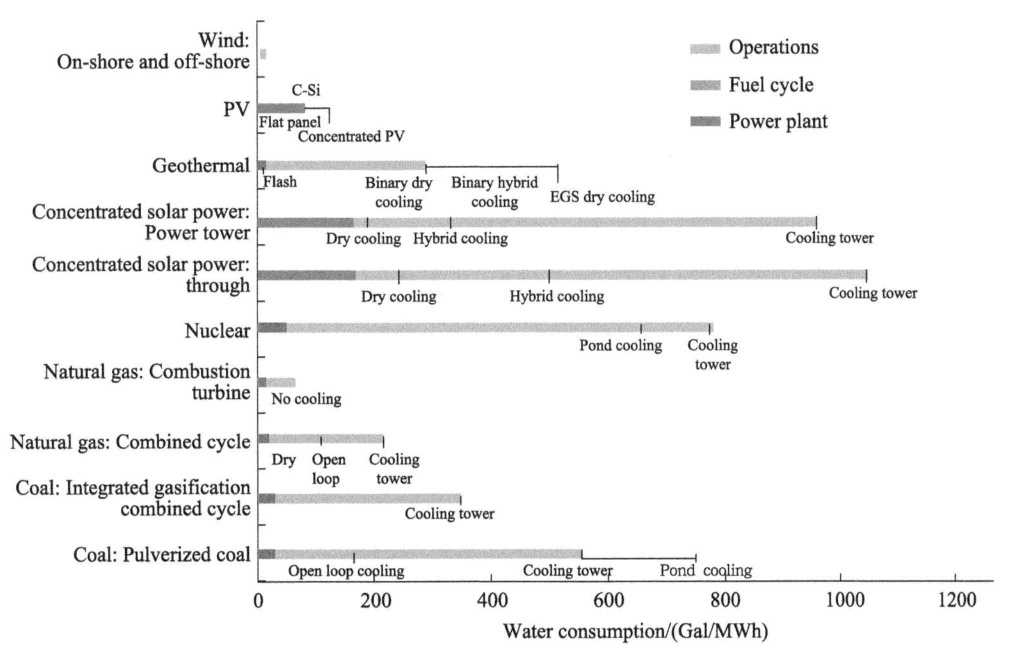


Figure 2 - 8 Water consumption by technology (Gal/MWh)

This figure shows that wind and geothermal plants consume much less water than nuclear plants. Many concentrated solar plants also consume less water, but this depends on

① As usual for water assessments, this report classifies water use into water withdrawals, referring to "water removed from the ground or diverted from a surface-water source for use", and water consumption, referring to the portion of withdrawn water not returned to the "immediate water environment".

their cooling schemes. This figure also shows that nuclear plants consume more water than gas and coal fired plants because of their lower thermodynamic efficiency.

The range of water withdrawals is also shown in Figure 2 - 9 (open cycles use an amount of water per MWh that exceeds the scale adopted in this sketch, and they cannot be fully displayed).

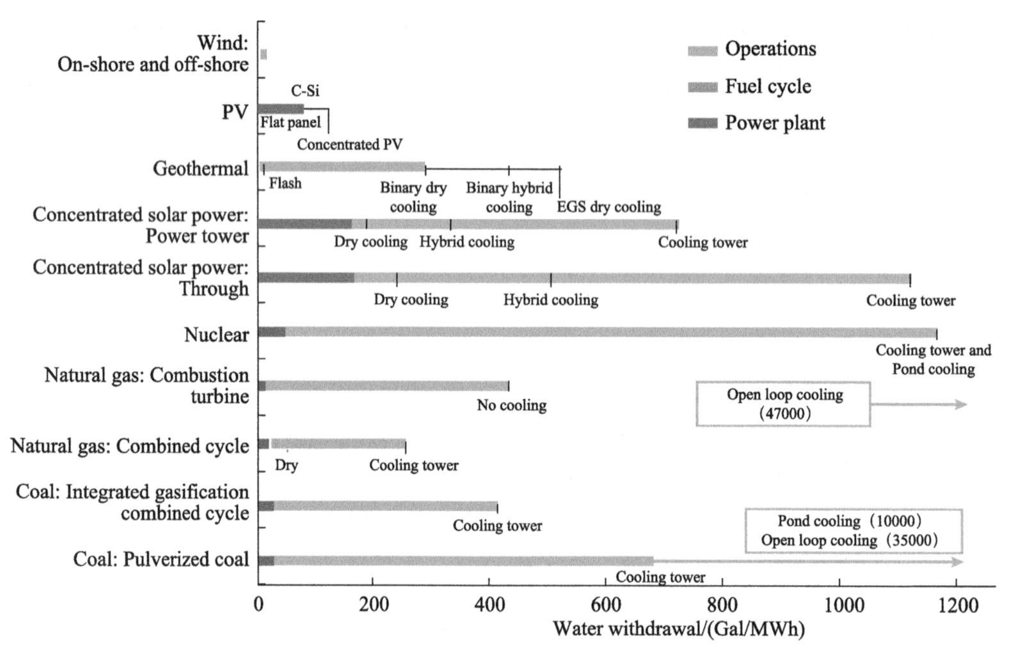


Figure 2 - 9 Water withdrawal by technology (Gal/MWh)

Typically, in France, water withdrawals for electricity generation account for 61% of water annually withdrawn (Table 2 - 4); but most of it is used by only six units having once through cooling and return to the river; the amount of consumed water calculated with the above figures which are applicable to all PWR units, is about 6% of the total annual withdrawals.

**Table 2 - 4 Source: INSEE/BNPE water statistics**

| Water withdrawal by sector in France/Mm$^3$ | 2013 | |
|---|---|---|
| Potable water | 5283 | 19% |
| Industry and other commercial usage | 2745 | 10% |
| Agriculture | 2776 | 10% |
| Thermal power generation | 17023 | 61% |
| Total | 27827 | 100% |

It holds true that a significant quantity of energy is dumped to the environment. In France, only its first six river-cooled units are using direct cooling; and authorities imposed wet tower cooling for all subsequent units; as the cooling water temperature is higher, it results in a loss of efficiency, and therefore of electrical generation, of about 4%. Furthermore, restrictions on the temperature of discharged water are implemented, which impose to reduce the plant output if constraints to the environment are excessive; however, their consequences remain limited: from 2000 to 2017, the average loss of output due to thermal constraints has been 0.18%, with a maximum of 1.2% in 2003 in a situation of an exceptionally hot summer (Ref. [19]).

Along the French Rhône river are sited 14 NPPs; their contribution to water temperature increase is 1.2 ℃ (average), or 1.6 ℃ (eighteen hottest days of the year), which remains reasonable (Ref. [20]). But the siting of inland power plants must be very carefully planned, as cooling the power stations may be competing with other needs, especially in water stressed regions.

### 2.3.4 Conventional waste from decommissioning

This paragraph does not deal with the radioactive waste that is treated in Chapter 3 of this report. But it is worth recalling that nuclear energy produces 4 and 3 orders of magnitude less technological non-radioactive waste than coal and oil, respectively. The graph "Material required" (Figure 2 - 6) shows that, except for lead, solar and wind power plants are consuming between 10 and a few hundred times more materials than NPPs. Amplified by the relative power factor of renewable versus nuclear, one needs twenty to a few hundreds more materials, such as concrete, copper and aluminum, per kWh in comparison to NPPs.

The relatively short service life of both solar and wind technologies should also be mentioned. Dismantling and reconstruction are relatively frequent and the recycling of at least part of the materials is an open question.

### 2.3.5 Critical materials

Another important parameter in terms of environment is the scarcity of some materials mainly used in solar PV and wind technologies (e.g. rare earths elements) vs. almost no use in the hydro and nuclear technologies (Ref. [15]). There is an exception with Nickel which can be considered as a strategic element: nuclear plants mobilize large quantities of stainless steel, and therefore of nickel.

## 2.4 New technology perspectives

Commitments taken as part of the Paris Agreement include the goal to "Achieving a balance between anthropogenic emissions by sources and removals by sinks of greenhouse gases in the second half of this century" (Article 4-2015). This goal is generally referred to as "Carbon Neutrality". The bulk of anthropogenic $CO_2$ emissions (87%) stems from the burning of fossil fuels like coal, natural gas and oil (Ref. [21]). Electricity, heat generation, and mobility accounted for 70% of the $CO_2$ emissions from burning fuels in 2014 (Ref. [22]). It is consequently of paramount importance to drastically limit, and if possible, replace fossil energy used in these sectors. In this section, we review the major perspectives to reduce carbon emissions from burning fuels. As one of the main means to achieve this goal is the use of nuclear energy, we also review the perspectives of reducing waste from nuclear power stations.

### 2.4.1 Reducing Carbon emissions of burnt fuels

Two main routes can be considered to reduce carbon emissions of burnt fuels: carbon capture and storage (CCS), and direct generation of carbon free electricity.

(1) CCS faces three challenges: reducing costs, improving public acceptance and developing storage capacities. Presently only 40 Mt/a of $CO_2$ are stored in the world;

4000 Mt/a should be captured and stored by 2040 (30% in OECD countries and 70% in non-OECD countries) according to IEA 2 ℃ scenario (Ref. [23]). It is also worth underlining that the efficiency of carbon capture is at the maximum of about 90%; therefore, it would not meet the goal of carbon neutrality without developing carbon sinks. All in all, carbon storage can only reduce $CO_2$ emissions, but won't be enough to achieve Carbon Neutrality.

(2) Carbon free electricity generation is the key to both a decarbonized mobility and a decarbonized energy system. Carbon free mobility can be provided by biofuels, or electricity stored in batteries, or generated by fuel cells converting $H_2$ into electricity.

Biofuels are one of the means to achieve Carbon Neutrality; buttheir limitations are acknowledged, as the energy per square meter of land which can be harnessed from biofuels is small compared to what solar or wind can provide for an identical land use. And they compete with the production of food which may have to be given priority when world population is steadily increasing. Therefore, leading technologies being considered to generate carbon free electricity are wind and solar. Their costs have plummeted in recent years; but they share the same limits of intermittency.

The only solution to cope with intermittency is storage. So far, batteries can be used for daily storage. However, much greater capacities than daily storage need to be considered. Batteries cannot be a solution to store thehuge amount of energy required to balance surplus and deficits over weeks or even months. Many alternate solutions could be considered: mechanical (compressed air, hydroelectric energy storage); thermal (molten salts, etc.). However, none offers the storage capacities which will be required.

From this review, it can be concluded that only hydro-for which available sites are scarce-, and nuclear have the potential to generate dispatchable and carbon free electricity.

### 2.4.2 Transmutation technologies

Transmutation is an option for waste minimization of HLW generated in the nuclear fuel cycle. The two main transmutation technologies are accelerator driven system (ADS) transmutation and fast neutron reactor (FRs) transmutation.

Scientists including Nobel laureate Carlo Rubbia promoted concepts of accelerator driven systems (ADS). In such systems, criticality would be achieved by the addition of an external source of protons, generated by spallation and accelerated, which would transmute fission products. Technical challenges faced by these technologies are significant, and their economic competitiveness for electricity generation is questionable. While they would have the potential to transmute actinides, fission products transmutation would be highly challenging. As the long-term risk of a geologic repository is usually dominated by fission products which are generally more mobile than actinides, the benefit of ADS-should their cycle efficiently work-would remain limited (Ref. [24]).

A fleet of FRs would essentially burn depleted uranium, circumventing the front-end of the fuel cycle, in particular uranium ore mining, thus further reducing the environmental footprint of nuclear energy systems. In addition, they could control the plutonium stockpile, minimizing the risk of its dissemination, and subsequent proliferation. The Generation-Ⅳ International Forum (GIF), a framework for international cooperation in research and development for the next generation of nuclear energy systems, encouraged the development of six promising reactor technologies, four of them being fast neutron reactors (gas-cooled fast reactor (GFR), lead-cooled fast reactor (LFR), molten salt reactor (including fast spectrum MSRs (MSFRs) ), sodium-cooled fast reactor (SFR) ).

Spent nuclear fuel back-end cycle has limited impact compared to front-end activities (ore mining and milling, conversion of $U_3O_8$ into $UF_6$, enrichment of $UF_6$, conversion of $UF_6$ to oxide). The lowest impact is provided by multiple recycling, and Fast Reactors.

Table 2-5 below is a comparison (Ref. [25]) based on a life cycle analysis of the French nuclear installed base, between:

(1) Once through fuel cycle (OTC) (spent fuel is considered as an ultimate

waste);

(2) Twice through fuel cycle (TTC) (spent fuel is processed once to recycle plutonium in MOX fuel and uranium in URE fuel, as deployed today in France);

(3) Gen-Ⅳ fast neutrons reactors fuel cycle (theoretical 100% sodium fast reactors design study but easily extendable to other Gen-Ⅳ fast neutrons reactor designs).

**Table 2 - 5 Comparison of three fuel cycle options**

| Impact indicators | Unit | OTC | TTC | SFR |
|---|---|---|---|---|
| $CO_2$ emissions | g/kWh | 5. 45 | 5. 29 | 2. 33 |
| $SO_x$ emissions | g/MWh | 18. 73 | 16. 28 | 0. 59 |
| $NO_x$ emissions | g/MWh | 29. 01 | 25. 3 | 3. 83 |
| Land use | $m^2$/GWh | 222. 6 | 211 | 50. 2 |
| Liquid chemical effluents | kg/GWh | 333. 92 | 287. 53 | 12. 6 |
| Gaseous radioactive release | MBq/kWh | 0. 8 | 1. 22 | 0. 53 |
| Liquid radioactive release | kBq/kWh | 2. 8 | 27. 2 | 3. 56 |
| High level waste (HLW) | $m^3$/TWh | 1. 17 | 0. 36 | 0. 3 |

It clearly demonstrates that multi-recycling activities improve environmental indicators.

The benefit of a linear combination of both 3rd (Gen-Ⅲ) and 4th (Gen-Ⅳ) generation reactors can be derived from this analysis.

Implementation of recycling substantially reduces the volume of high-level waste, which determines the size of geological repositories required by its high residual thermal power: with recycling, the repository volume and surface are divided by a factor greater than two.

The increase in radioactive gaseous and liquid releases with TTC compared to OTC results from dissolving the spent fuel in the reprocessing plant and is mainly due to krypton ($^{85}Kr$) and tritium. These radioactive releases are well below regulatory limits, and have a negligible effect on health and the environment. Their impact is lower than 10 μSv/a or 1% of natural sources of radiation except radon exposure.

### A short outlook on technologies being developed

Pool type, sodium cooled reactors remain the preferred route of development of fast reactors despite the many issues raised by this technology. Among ongoing achievements and developments, it can be mentioned:

(1) BN-800 reactor is a sodium-cooled fast breeder reactor, built at the Beloyarsk Nuclear Power Station, in Zarechny, Sverdlovsk Oblast, Russia, achieving commercial operation in 2016.

(2) China has plans to develop a 600 MWe demonstration SFR based on the CEFR (China Experimental Fast Reactor-65 MWth-20 MWe) experience.

(3) The French Atomic Energy Commission (CEA) completed the basic design of ASTRID (advanced sodium technological reactor for industrial demonstration), a 600 MW sodium-cooled reactor including an advanced sodium gas concept to transfer energy from the reactor to a gas turbine.

(4) Other technologies are also being considered. Among them, it is worth to mention molten salt fast reactors concepts, such as the MOSART concept in Russia without or with Th-U support, or the Molten Salt Fast Reactor, sketched by the French Research Institute CNRS. These reactors claim to allow a progressive shift from Uranium based (scarce) to Thorium based (abundant) cycle if needed. However, this goal may be questioned, as the Fast reactors by themselves would alleviate any concerns related to Uranium scarcity; and use of Thorium would require investing in a completely new fuel cycle infrastructure.

## 2.4.3 Other advanced technologies

Other advanced technologies are:

(1) Development of new generation of accident tolerant fuels (ATFs); ATFs are designed to better withstand a nuclear accident; SiC fuel cladding would be beneficial for fuel cooling under normal operation, and limit fuel temperature by notably reducing hydrogen generation from zirconium water reaction with chromium coating or SiC

cladding. Fuel pellet would be designed to increase thermal conductivity and reduce radioactive release. By themselves, ATF have no impact on waste.

(2) Artificial intelligence (AI) tools, protected against cyber attacks, will help operation of NPPs, combining sensors integrated to NPP equipment and NPP numerical twins with algorithms delivering diagnosis and monitoring equipment behavior.

## 2.5 Conclusions

As a general conclusion, one may note that impacts of nuclear energy on the environment are well documented.

Concentrations of radionuclides in the environment are easy to measure and counter-check. It is more difficult to quantify concentrations of chemotoxic elements.

The monitoring of radioactivity in real time is a warning sign to preserve the environment and the data obtained allow checking the results of simulations of real or expected releases. The controversial problem, if any, is the estimation of the associated detriment. For radiological detriment, the dose is an additive parameter. Such parameter does not exist in the protection against chemotoxic substances.

Nuclear energy does not release chemical pollutants resulting from combustion to the atmosphere (Ref. [5]). The impacts on the environment come from the potential release of radioactivity at each stage of the nuclear fuel cycle, for instance during transportation.

The radioactivity levels of the releases are regulated by the competent radiation protection authority. Discharge authorizations are provided on the basis of doses to the most exposed individuals or critical group close to the site, according to scenarios. The actual releases reach only a few percent of the authorized levels. There are also limitations based on the calculated maximum admissible limits in Bq fixed by international organizations for waters, air and some bio-indicators. They derive from scenarios taking into account exposures due to all possible radioactive releases and corresponding to a committed maximum dose (see later) of 1 mSv in one year.

The local radiological impacts on living species, including humans, are usually low to very low, when compared to those due to the radioactivity of natural environments or to that associated with the use of fossil fuels like coal or shale gas; as a matter of fact, they are so low that they cannot be identified. It is the same situation for chemical pollutants released from nuclear facilities compared to other possible transfers from man-made activities. These assessments are based on the results of numerous monitoring devices implemented by operators, authorities and stakeholders, and of epidemiologic studies.

According to LCA of energy systems, which take into account all impacts on the environment generated during the fabrication of materials for the construction of power plants and facilities as well as during their operations, nuclear energy is characterized by a quasi-zero emission of $CO_2$ (and other greenhouse gases) compared to all fossil fuel-fired energy systems and is the energy that is consuming the least amount of land and material. With regards to other decarbonized sources of energy, nuclear energy occupies in one to two orders of magnitude less land for the same amount of energy production, needs less concrete and steel per MWh and does not demand critical materials such as rare earth elements.

The consumption of water to cool inland reactors is significant, and must be seriously considered in water stressed regions. However, there are no significant cooling water issues for plants cooled by sea water or large rivers.

If one focuses on civil works to build a Gen-Ⅲ nuclear reactor, it appears that the use of concrete and steel is higher than for a fossil fuel-fired power plant. This is due to safety requirements with regard to external and internal events (earthquakes, aircraft crash, etc.).

Fast reactors and multi-recycling have the potential to drastically reduce the environmental footprint of nuclear energy, and their developments should be adequately funded.

# Appendix 2 - 1 More specific considerations about the Chinese situation

The growth of nuclear energy in China is the fastest of all nuclear countries. In recent years, China has carried out much research on greenhouse gas emissions and the environmental impact of radiation from nuclear power, fossil energy (coal-fired power stations), and renewable energy sources (hydropower, wind power and photovoltaic power generation), using life cycle analysis (LCA) to examine the direct greenhouse gas emissions and radioactive releases during construction and operation, as well as the indirect emissions and releases from energy and raw material consumptions of the energy systems and related infrastructures during mining, manufacturing, processing and transportation. The emissions from main materials used for construction are added to the total emission of the power source.

Figure A2 - 1 shows the normalized life cycle emissions of greenhouse gases from different power sources: the greenhouse gas emissions from nuclear power, hydro-

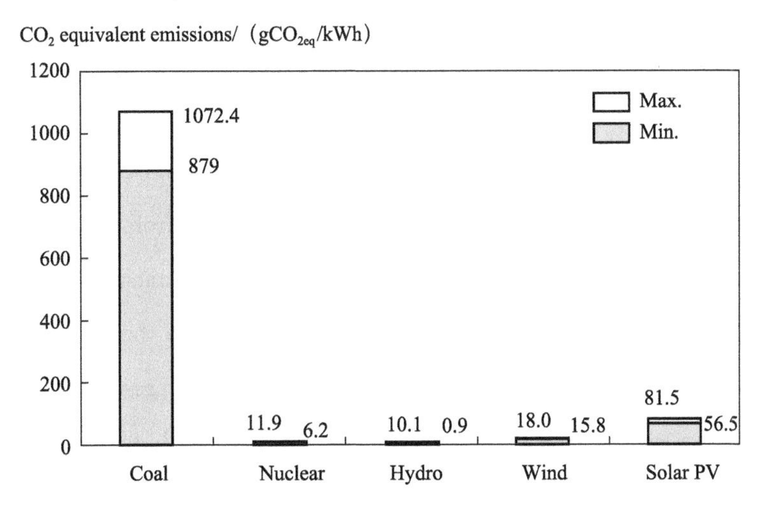


Figure A2 - 1 Normalized greenhouse gas emission from different power sources in the life cycle (Ref. [26])

power and wind power are lower than those from coal power by two orders of magnitude. The emissions from photovoltaic power generation are lower than those from coal power by one order of magnitude, placing them at the medium level. For the nuclear power aspect, greenhouse gases emission due to energy consumption account for 84% of the total emission, indicating that emission related to nuclear power depend on the Chinese energy mix. If the primary power is provided by NPPs or other renewable energy sources instead of coal power, every generation of 1 kWh power could decrease $CO_2$ emissions by 1 kg, featuring considerable potential for the reduction of greenhouse gases. It is predicted that, from the year of 2020 up to 2050, the share of non-coal-fired power generation in China will increase from 28% to 47%, and the share of coal power will diminish from 69% to 49%. Furthermore, in that case, when power-generating capacity in China grows by 70%, the emission of greenhouse gases will rise by merely 23% (according to the maximum level of normalized greenhouse gas emissions from different power sources). In conclusion, clean energy sources (nuclear power and renewable energy sources) have great potential capacity for the reduction of greenhouse gas emission, and are essential to build a low-carbon energy system and speed up the transformation of power generation and demand.

Figure A2 - 2 shows the normalized collective dose from different power sources in the life cycle (assessment range: 80 km distance to the site). As for hydropower, wind power and photovoltaic power, the public dose is relatively low (until today, there is no such research on the direct radioactive releases from the development and electricity generation of renewable power sources, such as for example radon released from water in hydropower plants). As for NPPs, 86% of the public dose comes from the mining of uranium, but as the *in-situ* leaching uranium mining technologies get expanded in the near future in China, the public dose from nuclear power will be further reduced. At present, coal power is the first and foremost energy in the Chinese energy structure. With the upgrading of power plant structures, making the 300 MWe and above units gradually becoming mainstream, and the reduction of coal consumption for power supply together with the development of dedusting technologies, public doses generated from coal power (except for the utilization of coal ash and slag) will drop greatly. But coal has relatively low energy density and generates large amounts of ash and slag. In China, such ash and slag are generally mixed to be used as the main material for housing walls, generating $2.6\times10^3$ man • Sv/GWa of normalized collective dose to the public (average from the year of 2003 to 2010), accounting for 99.9% of normalized collective dose to the public in the life cycle of coal power, higher than the sum of others by nearly three orders of magnitude and considerably higher than that of other power sources. Thus, research results indicate that improving the energy structure and developing nuclear and renewable power could greatly reduce the public exposure.

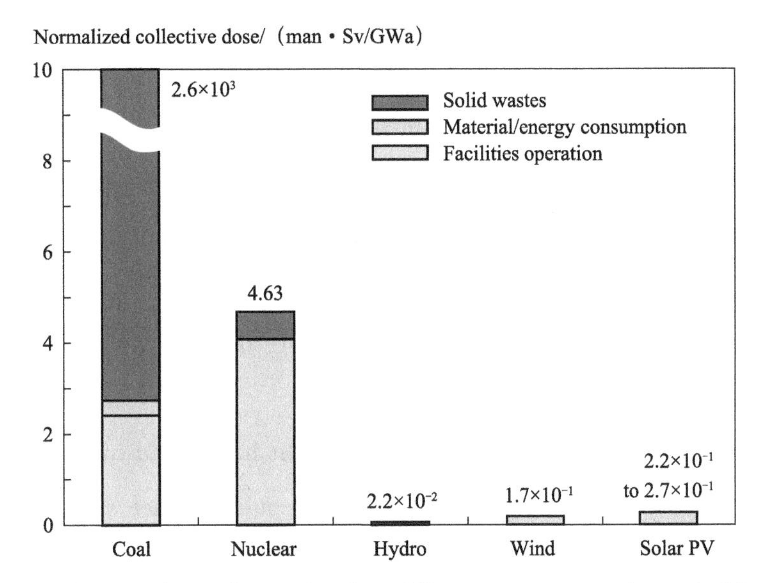


Figure A2 - 2 Public normalized collective dose from different power sources during the life cycle in China

Note: In terms of coal power, "Solid waste" refers to the application of coal ash and slag on the main walls of houses; in terms of nuclear power, "Solid waste" refers to the disposal of solid waste

# Appendix 2 - 2 Large scale epidemiological studies around nuclear sites (cases and results)

As releases of radioactivity around nuclear facilities are quite low, their impacts on life and health of the surrounding populations can be evaluated only by means of epidemiological studies, which require probabilistic assessments. They analyze the cases presented or the mortality rate caused by radioactivity induced diseases, as well as the factors that influence their development. The studies are done in the natural environment of the observed individuals, and their life habits are taken into consideration as far as possible.

Since the World War Ⅱ, hundreds of epidemiological studies have been carried out within the environments of nuclear sites and contaminated areas all over the world.

Conducting epidemiological studies is not easy. This requires a good preparation, a precise identification of all the interfering parameters and their interrelations, a detailed and in-depth study of the environmental conditions that might influence the research, a prolonged observation time, orderly and thorough data collection, adequate means and properly trained specialists to correctly interpret and handle the data, the methods used, and the results obtained.

Here are some of the reasons which complicate the analysis of radiation consequences at low doses:

(1) Given theirlow incidence, the effects might be masked by causes other than ionizing radiation, which, at greater frequencies, could produce similar effects in an isolated or simultaneous way.

(2) From the methodological and statistical viewpoint, and due to this low incidence, it is necessary to study very large population samples throughout several generations and along with very large contrasted population groups (control samples) with similar environmental factors and which have not been exposed to the ionizing radiations.

(3) Humans are being continuously subjected to natural ionizing rad iation (such as cosmic radiation and radiation from radioelements in the atmosphere and in the earth's crust) as well as from artificial radiation (medical or industrial uses of nuclear radiations) or non-ionizing soft radiation (such as radiation from television, computers …). Therefore, it is not easy at all to discern the effects produced by one

or the other source of radiation.

In this respect, one may mention a few international epidemiological studies:

(1) The 2006 to 2010 study of "Possible Radiological Impacts of Nuclear and Radioactive Sites on Human Health" carried out in Spain by the Ministry of Science and Innovation, the Carlos Ⅲ Health Institute, and the Nuclear Safety Council. *This study concluded that nuclear sites do not affect the risk of cancer for the population; the estimated accumulated doses that the population would have received in the analyzed areas are very low, on the average of what is about* 1/300 *of the natural background radiation around the sites.*

(2) The 2008 to 2010 study carried out by the University of Berne in Switzerland at the requests of the Swiss medical *authorities showed that, there is no relationship between juvenile cancer and Swiss nuclear reactor sites.*

(3) In 2008, the French IRSN published a detailed report (Ref. [27]) which synthesizes the results of all the epidemiological leukemia studies around nuclear sites (of all types) done in the world. The main conclusion is the following: *At the local level, excess of acknowledged childhood leukemia cases exists in the UK near the Sellafield and Dounreay reprocessing facilities and in Germany near the Kruemmel NPP. Nevertheless, all the multisite studies available today, including France, do not show an increase of the frequency of leukemia among young people* (0 - 14 *years or* 0 - 24 *years*) *around the nuclear sites.*

(4) A German study shows an excess of leukemia among the (0 - 4 years) around 16 German NPP (Ref. [28]); but the authors caution the readers that their findings are unexpected given the very low observed levels of radiation and they state that the cause of childhood leukemia remains unexplained and may be due to uncontrolled confounding or pure coincidence. Today, such an observation is not confirmed by the studies carried out in other countries, including France.

Numerous studies have tried to explain the excess of leukemia observed around some nuclear sites by looking at multiple potential risk factors. But the determination of the causes is limited by the lack of knowledge about risk factors of childhood leukemia, especially on potential effects of ionizing radiation exposure in utero and during early childhood. Large scale investigations would be necessary, at national and international levels.

Epidemiological studies are often done once a nuclear facility is already in place. They cannot reveal changes if there is no reference to the situation prior to the construction of the nuclear facility. Therefore, one would need studies before and after the installation of a nuclear facility.

It appears that, in order to study health effects, epidemiologic studies around nuclear sites are more appropriate when carried out after releases of radioactivity during accidents than regular investigations, which cannot easily reveal the very small impact of facilities during normal operation. Nevertheless, the latter are interesting as background for comparisons.

# Chapter 3 Spent fuel and radwaste management

---

Recommendations

The present analysis indicates that radwaste management under present practices has very low local and global impacts on health and the environment. Nevertheless, improvements of the various processes, which lead to the release of radioactivity outside of nuclear facilities and radwaste packages, are always desirable.

The Academies recommend that the methodology to evaluate all environmental impacts (radiological and chemical) and associated risks should be refined considering waste arising from the front and back-end of nuclear fuel cycle and taking into account time scales.

To support this general recommendation, the Academies propose that:

(1) The best available technologies (BATs) should be used to confine radionuclides at every step of the processes;

(2) The process of safe disposal of radwaste should be accelerated, so as to ensure intergenerational equity and to avoid undue burdens on future generations;

(3) R&D programs should be developed aimed at a better understanding of the radiological and chemical impacts on ecosystems (reversibility, resilience, bio-availability of elements of interest …) and quantitative parameters should be defined to characterize the hazards linked to radwaste in order to better meet environ mental issues;

(4) Comprehensive and responsible systems, visible to the public, should be used to protect the environment (including legislation system, competent bodies, funding system …).

---

## Introduction

A specificity of the nuclear industry is that it uses fuel that does not disappear when "burned" . The nuclear industry cannot manage waste in the same way as the fossil fuel industry does, i. e. according to standard waste disposal channels in the form of greenhouse gas emissions into the atmosphere on one hand and accumulation of solid residue deposits on the other. Fission and other nuclear processes inside the nuclear fuel produce around a hundred of short- or long-lived radionuclides, i. e. radioactive isotopes which encompass two thirds of the elements of the periodic table. The chemical properties of all these radionuclides are drastically different. The radioactivity of nuclear fuel increases, from a $kBq/cm^3$ (fresh fuel) to $10^{10}$ or $10^{11}$ $Bq/cm^3$, when it is downloaded from reactors (spent fuel). All electronuclear radwaste contains higher or lower amounts of these radionuclides. Management of radwaste is then a part of the nuclear fuel cycle. Today industrial channels for radwaste management are operated in all nuclear countries. The great majority of radwaste (the less radioactive and more abundant) is finally disposed of in surface/sub-surface repositories; the remainder (the more radioactive and less abundant) is kept in storage pending the launching of deep geological repositories. Despite of the high level of care taken in such operations, sorting out fissionable material is still held in spent fuel (essentially plutonium), and nuclear waste leads to the immediate release of a few radionuclides into the environment. In a long-term future (from centuries up to thousands of centuries) one may expect the return to the biosphere of some radionuclides from disposed-of radwaste in the geosphere. However, in all cases, measures are taken today to keep the radiological impacts within the normal variations of natural sources of radiation irrespective of geography and timescale.

This Chapter focuses on the environmental impacts associated with the management of radwaste.

# 3.1 Principles, strategies and framework of radwaste management to prevent environmental impacts

The following principles and strategies reflect the international situation but are mainly drawn from several decades of French return of experience.

## 3.1.1 Principles of radwaste management

The first basic principle is the inter-generational equity (i.e., our generation should not leave the burden of our technical decisions to future generations). The environment is the common property of all generations. Leaving a clean environment to the next generations is a major duty of the present one, particularly with respect to restraining addition of radioactivity to the natural radioactivity. The second is the inter-generational right of access to information so that each generation remains informed about the practices of radwaste management at national and international levels. Keeping the memory, as long as possible, of the location of radwaste having possibly an impact on the environment is the duty of national and international organizations in charge of radwaste management.

To restrain releases of radionuclides to the environment, operators have to implement the BATs (best available technologies) for radwaste management in all nuclear facilities, to minimize radwaste production. This is already the current practice as described in the following.

## 3.1.2 Strategies of radwaste management

The global strategy defining radwaste management is to ① maximize in-reactor burning of radioactive materials, ② concentrate and confine radionuclides and toxics and ③ finally dispose of ultimate radwaste in repositories. These engineered infrastructures are designedto isolate radwaste from the biosphere in such a way that the time of return of radionuclides to the living world would be as far-off as possible: in terms of centuries or thousands of centuries. Dilution strategy is avoided. Thus, radwaste management basically differs from the management of conventional waste. It requires a high level of scientific and technical competence and strong support. The principle of "Safety First" should be implemented, even if at the detriment of economy.

Except for radwaste produced in very large quantities such as in uranium mining/milling or low/very low level radwaste, crude waste produced at each stage of the fuel cycle is-as soon as possible-either processed to confine/isolate radionuclides or stored in facilities to prevent any contact with the public and the environment, waiting for further processing. The objective is to produce primary waste packages that are handled in all the steps leading to their final disposal. Packages used for high-level waste rely on the best technology design concepts either for conditioning fission products and minor actinides when spent fuel is reprocessed or for encapsulating spent fuel assemblies when spent fuel is not reprocessed (See Section 2. 1). Radionuclides cannot escape from these packages (except under highly hypothetical circumstances). For other radioactive waste, whose radioactivity is lower by several orders of magnitude, packages are not necessarily sealed.

When repositories are not immediately available, the packages of whatever radwaste are stored in specially designed facilities, pending final disposal. For high-level radwaste packages, storage in such facilities is mandatory to allow a decrease of their thermal power before being disposed-of.

Short/medium-term impacts to the environment originate from processing and packaging. Under normal operation of the facilities where crude waste is processed and where the packages are stored, the releases to the environment are below the authorized limits as is the case for all nuclear facilities. Long-term impacts come from packages that have been disposed-of, because a time comes when the containers corrode, and radionuclides/toxics are slowly released. During uranium mining and disposal of non-packaged radwaste from processing of uranium ores, both types of impacts (short- and long-term) have to be considered.

Radwaste is disposed of in several ways including surface, subsurface, or will be disposed of in deep geological repositories, depending on their characteristics. Whatever they are, the decision to open a repository, which is a nuclear facility, relies on a safety case analysis, requiring preservation of public health and avoidance of environmental pollution. This safety case analysis considers both short/medium and long-term impacts.

For short/medium term in all circumstances, hazardous impacts on people and the environment must be well below legal thresholds and in compliance with current regulations.

The long-term, potential impacts are assessed by making use of simulation based on the present up-to-date information, data and scenarios. Discharges of radioactive gaseous or liquid effluents to the environment lead to the dispersion of radioactivity and to their deposition at a more or less distant location from their emission source. Finally, radionuclides enter the biogeochemical cycles. The immediate radiological and chemotoxic impacts of these fall-outs can be calculated from *in-situ* measurements. This is not the case for the long and very long-term impacts of radionuclides returning to the biosphere from contaminated soils or waters or from radwaste packages disposed-of in geological formations. Scientists have to model the space-time migration of radionuclides through many natural or exogenous materials from the place of their emission to the outlets of the formations before to estimate the impacts through scenarios. Time duration to be considered in safety case analysis goes from ten thousand years for quantitative estimates of doses to a million years for qualitative estimates, well beyond usual theoretical and practical considerations in the field of technology and stability of society.

The long-term modeling depends on the understanding of effects due to the complexity of the microscopic and macroscopic convection or diffusion phenomena. Convection phenomena occur in fractured engineered or geological materials. The transport of solutes in large homogenous materials without connected fractures is under the dependence of concentration gradients leading to theoretical diffusion laws in $t^{1/2}$. The behavioral laws of the materials used for their ability to withstand degradation like packages of high level/long life waste are, in the final analysis, the result of microscopic diffusion phenomena. In total the long-term models incorporate low-power empirical laws in relation to time: $t^n$, $n$ being less than 1, meaning that consequences are decreasing over time. Sciences which support the models belong to the fields of earth science, material science and life science. Each field has accumulated many data.

Earth sciences are familiar with the mechanisms of thermal, hydrogeological, mechanical and chemical (THMC) evolutions of geological layers up to a million years and have local models for earthquakes, hydrogeology and climate. The astronomical cyclical changes that affect the climate are included as part of a global climate model. Future events and their consequences can be included on these bases over a few hundred thousand years.

Material sciences can model the fate of rocks, solid components of waste packages and storage structures, taking into account all knowledge derived from natural or anthropogenic analogues and experimentation. The same is true for the migration of elements. The importance of geochemistry is of a crucial importance. Thus, a deep reducing environment ($E_h \ll 0$) allows asubstantial limitation of the solubility of actinides and thus of their potential migration. This is also true for many fission products. The only long-lived radio-elements that are not susceptible to redox conditions are essentially $^{36}Cl$ or $^{129}I$, which largely dominate the outfall inventories predicted in normal and degraded scenarios. The majority of radionuclides remains trapped on the spot in the near field.

Models are constructed on the basis of experiments carried out over decades and results of brutal, mild, repeated disturbances aimed at accelerating environmental aggressions. The results obtained generally allow for the identification of mechanisms at the appropriate scale. The models can be tested by simulation and blind validation in which information is initially masked. Uncertainties can be reduced by repeating the experiments. The limits of extrapolation in the different domains are known.

In material science, temporal modeling can be based on well-founded mathematical models. In life sciences modeling is less easy because of the complexity of metabolism. In human sciences, the historians work over a few centuries while sociologists consider a few decades, and the use of formalization and logic are less systematic, so that temporal models are, less advanced, or non-existent.

There are some uncertainties about future environmental impacts as in all cases where long-term simulations are being considered, mainly because scenarios require many assumptions. However, all simulations, even those corresponding to the worst scenarios, indicate that radiological impacts are well below the impacts of natural radioactivity (or of the same order in the case of human intrusion in the disposal).

### 3.1.3 Frameworks of radwaste management with respect to the environment

The frameworks of radwaste management are defined at the international level by:

(1) The Common Convention (IAEA, INFCIRC/546, December 24, 1997) on the safety of management of spent fuel and radioactive waste. This Convention is the result of broad discussions between 1994 and 1997 following the *Convention on Nuclear Safety* (IAEA, INFCIRC/449, July 15, 1994). It includes a section on protection of the environment against ionizing radiations. It calls for periodic reports from contracting nuclear countries about how they accomplish their obligations with regard to the dispositions of the Convention. Today 43 countries report each three years to IAEA.

(2) The recommendations of ICRP (IAEA, Safety series 115-I, Vienna, 1994) which are taken into account by all countries.

European countries have also to consider the European Council Directive 2011/70/Euratom which requires that each country develop a waste management policy protecting human beings and the environment.

In addition, there are, under the IAEA authority (and European Commission), international regulations dealing with the transportation of radwaste packages which require robust canisters to protect the public and the environment.

The implementation of the commitments of the Common Convention is a key item. This is why the IAEA follows the three years update of the contracting parties.

Both China and France are parties to the Common Convention on the safety of the management of spent fuel and radwaste, and they have established a relatively comprehensive legal and organizational framework for the effective and safe management of spent fuel and radwaste.

The French Constitution acknowledges the precautionary principle. This applies to all activities and aims at protecting the environment. It implies to take proportionate measures as soon as there is a presumption of serious and irreversible environmental impact. It involves new researches to better understand the fear phenomena (Section 5 of the 2004 *Environment Charter*, incorporated in the French Constitution).

In France, the management of radioactive waste is governed by two laws: the 1991 Law (focused on researches) and the 2016 Law (focused on implementation of decisions). The Ministry of Ecology Transition and Solidarity develops the policy and implements the government's decisions. Several nuclear operators are involved: EDF which operates 58 nuclear reactors, Orano-Cycle and Framatome which operate the facilities of the front and back-end of the nuclear cycle, CEA which leads R&D on nuclear energy and Andra which is in charge of the long-term management of radwaste. The safety authority ASN assures, on behalf of the French state, the control of nuclear safety and radiation protection so that people and environment are protected from risks related to nuclear activities. IRSN is the technical support of ASN for safety case analysis. It carries out its own researches on the noxiousness of radwaste and on environmental impacts of radwaste management. Finally, there is a high committee for transparency and information on nuclear safety. This committee formulates recommendations to improve the transparency and the quality of information for the public.

In France, all researches and investigations dealing with radwaste management are indexed in a National Plan (PNGMDR—*Plan National de Gestion des Matières et*

*Déchets Radioactifs*) issued every three years by ASN and the Ministry of Ecology and of Solidarity in connection with a pluralistic working group including environmental protection associations. This constitutes a strategic tool to implement the waste management policy and to inform the public. It identifies the needs, sets the objectives, makes recommendations to all stakeholders and asks for mandatory reports. It is evaluated by a commission in charge of the environment. The Plan is transmitted to the Parliament and submitted to public consultation.

A complete inventory of all radwaste and nuclear materials is published by Andra every three years gathering the declarations of producers. It comprises radwaste categories, quantities, geographic location at the time, and for the next decades, according to several scenarios of evolution of the domestic nuclear fleet. Any change has implications on the nature and quantities of waste and possibly on the impacts on environment.

Each year an independent committee set up by law (CNE—Commission Nationale d'Evaluation) reports to the Parliament about results of research carried out on this topic.

In China, a legislative framework for governing the safety of spent fuel management and the safety of radwaste management is established and maintained, which incorporates a comprehensive set of relevant national laws, administrative regulations, departmental rules, management guides and reference documents, as well as the licensing regime of spent fuel and radwaste management activities. The laws applicable to the management of spent fuel and radioactive waste are: a) the *Law on Prevention and Control of Radioactive Pollution* (LPCRP), enacted by the NPCSC (National People's Congress Standing Committee) in 2003, and b) the *Act of Nuclear Safety*, enacted by the NPCSC in 2017.

In China, the Ministry of Ecology and Environment/National Nuclear Safety Administration (MEE/NNSA) is responsible for the regulatory control of the spent fuel and radwaste management, and the China Atomic Energy Authority (CAEA) is the competent body for the spent fuel and radwaste management.

In China, the generator of radwaste shall bear the overall safety responsibility for radwaste management and implement the management of radwaste in terms of their classification.

## 3.2 Specific characteristics and classification of radwaste

Nuclear countries have adapted the classification of nuclear radwaste to their national industrial channels and the ensuing management practices of radwaste categories may differ to some extent but there are many commonalities. The Academies have discussed this topic in their first report according to activity and life of radionuclides contained in radwaste. In the following, the traditional denominations used are as follows: very low-level waste (VLLW), low level-long lived waste (LL-LLW), low and intermediate level-short lived waste (LIL-SLW.), intermediate level-long lived waste (IL-LLW), and high level-long lived waste (HLW).

Before proceeding, it is mandatory to recall that the type of highest activity radwaste (IL-LLW and HLW) depends on the decision made by nuclear countries with regard to the spent fuel cycle.

### 3.2.1 Spent fuel or reprocessing radwaste

The management of radwaste from electronuclear reactors deals with:

(1) Spent fuel (when spent fuel is considered as radwaste) and radwaste originating from reprocessing and recycling of spent fuel (when spent fuel is considered as a source of fissile materials).

(2) All other radwaste produced by reactors or facilities of the nuclear fuel cycles.

As already indicated, there are two types of nuclear fuel cycles.

- Open nuclear fuel cycle, where the spent fuel is not reprocessed but stored under wet or dry conditions, pending the transfer to a final disposal after adequate conditioning and packaging as HLW packages of encapsulated spent fuel to be disposed-of are designed to be unfailing during thousands of years.

- Closed fuel cycle where the spent fuel is reprocessed, with valuable materials (U and Pu) extracted and the remaining material conditioned in the form of nuclear glasses and packaged mainly as HLW and IL-LLW. Packages of nuclear glasses to be disposed of are designed to be unfailing during thousands of years.

The radioactivity of spent fuel and nuclear glass is roughly the same, but the latter does not contain plutonium and uranium.

Each nuclear country in the world has chosen one or the other of these fuel cycle options, according to the national political, economic, technical or diplomatic contexts. France and China have chosen the closed fuel cycle scheme, allowing a sustainable nuclear energy policy. In Europe, the 28 countries of the European Union (including UK) have not all adopted the same strategy. Some countries consider that their choice is not final and could change with respect to the evolution of technology and return of experience of the more advanced nuclear countries in recycling fissile radionuclides; some countries have no electronuclear facility.

According to the characteristics of the packages of spent fuel and nuclear glasses to be disposed of, little difference is to be expected regarding the long-term environmental impact, if any. The same would apply to storage under monitoring in spite of the differences in the conditions of storage (sub-assemblies of spent fuel in pond or in dry cask, nuclear glasses in dry facilities). In contrast, reprocessing, which needs chemical separation of radioactive elements from spent fuel, has a local impact on the environment that is higher than the no-reprocessing option.

In this chapter, for the sake of simplicity, we shall generically consider the main trends in radwaste management independently of the nuclear fuel cycle option, with the objective of examining phenomena that could lead to environmental impacts.

### 3.2.2 Specific characteristics of radwaste versus the environment

Nuclear energy produces much smaller amounts of waste per MWh compared with fossil energies. This is linked to the energy density of nuclear fuel, thousands of times higher than that of fossil fuels, depending on burn-up of the nuclear fuel and on the reactor type. Radwaste is generated by a relatively small number of NPPs and fuel cycle facilities when compared to the gaseous and solid residue output of the many fossil fuel fired power plants. This condition makes radwaste management easy to control by Safety and Environmental authorities, in compliance with strict regulations, which are accepted by nuclear countries and derive as already indicated from recommendations of international agencies. Streams and characteristics of radwaste are the best documented of all waste streams. In those nuclear countries that have also legacy radwaste, produced by the early electronuclear reactors and by military applications, the corresponding stockpiles are well identified.

The main environmental impacts that one can expect from radwaste management are linked to public exposure to ionizing radiation and to modifications of the quality of aquatic and terrestrial ecosystems, possibly leading to a loss of biodiversity. They are due to liquid or gaseous releases of radioactive substances during severe accidents. Other impacts are those encountered in any supply-dependent operating facilities, involving important and continuous transportation of materials (resulting in heavy traffic, noise …).

The doses (external and internal) due to release of gases or liquids containing radioactive or toxic substances are estimated according to tested methods and the results are submitted to international scrutiny (round robin tests). Phenomena involved in the various processes are well understood and open databases are used to feed simulation models. Similar methods are used to estimate the impacts, if any, from toxic chemicals with reference to the recommendations of the World Health Organization (WHO).

It is less easy to quantify the impacts of radioactivity and toxics on other ecosystems because data on the non-human biosphere is still lacking. Usually, human beings are more sensitive to radiation hazards compared to other species. The non-human species would be suitably protected when human beings are adequately protected against radiation.

In both cases, the main knowledge-gap relates to the bioavailability of elements that carry radioactivity, i. e. the radioactive chemical species that can be transferred to living beings. In this respect, speciation of elements is of primordial importance. Radionuclides in the environment are at tracer scale and their physicochemical behavior cannot be inferred from the behavior of the element of which they belong at usual concentrations. Radionuclides species do not control chemical systems; on the contrary they undergo the constraints imposed by the species in weighable quantities. The chemical identity of the radioactive elements is lost. In principle, radioactive monomeric species could exist but they are often sorbed on natural colloids and appear as radioactive pseudo-colloids. In contrast, the behavior of radionuclides in packages is that of any element at usual concentration. Solubility phenomena limit their releases, a point that needs to be underlined. Many national and international research programmes have been set up to clarify basic phenomena involved in the transfer of radioactivity to living materials. This topic should be included in the numerous transdisciplinary investigations aimed at understanding the human impact on the environment. Ecotoxicology of radionuclides has been studied since several decades, but progresses are slow.

Other energy industries generate various non-radioactive waste containing toxic and hazardous materials. A special treatment and/or disposal in landfill could be required, but as the concentration of chemicals before having a toxic effect is much higher than that of radionuclides causing radiological hazards, confinement of conventional disposal sites for chemicals may be less effective than that of radwaste repositories. It is then apparent that radwaste repositories designed to confine radionuclides also effectively confine toxic material usually encountered in classical waste.

### 3.2.3 Classification of radwaste versus the environment

Spent nuclear fuel is periodically removed from NPPs and replaced by fresh fuel. As indicated, spent fuel assemblies are either reprocessed or stored before being disposed-of as high-level radwaste. For the reprocessing route, spent fuel rods are cut open, thus allowing the chemical separation of Pu and U through recycling. Today all other radionuclides resulting from recycling are dissolved in nuclear glasses and packaged. The resulting packages are initially as radioactive as the original spent fuel but with very small content of uranium and plutonium. The reprocessing and manufacturing processes generate additional streams of radwaste, more or less contaminated with long-lived radionuclides present in spent fuel. The operation of NPPs also generates radwaste. Finally, large quantities of various categories of radwaste, but mainly of low-level radioactivity, are expected from future massive decommissioning of reactors and facilities.

Regarding the environmental impact, a distinction between short-lived and long-lived waste is crucial. Indeed, the former is generally disposed-of in surface/sub-surface facilities and its environmental impacts can logically be of direct concern to our generation. Long-lived waste will be disposed-of in deep geological repositories, down to several hundred meters, and its expected environmental impacts are seen as possibly occurring in the far future. Notwithstanding, both strategies are a subject of careful investigation.

An additional distinction considers the origin of radionuclides present in radwaste: natural (U, Th and daughters) or man-made (actinides, fission products, tritium, etc.). Radwaste containing only uranium originates from the front-end nuclear cycle. Radwaste linked to the back-end of the nuclear cycle contains in addition many other additional radionuclides.

Releases of radionuclides or toxic substances can occur during the operation of the facilities/repositories and even after their closure. The operational procedures and equipment for dropping-off the packages as well as the daily and periodic monitoring of radioactivity (and sometimes of chemical pollutants) permit the detection of any malfunction affecting normal operation. Thus, local environmental impacts need to be evaluated according to the selected domains where they might occur. The occurrence and extent of releases of radionuclides and toxics, long time after the closure of repositories, depends on the robustness of packages and the capacity of manufactured and natural barriers to prevent migration of elements to the biosphere. As already indicated, near-and far-field impacts can only be modeled. In France the classification of radwaste follows the classical approach implemented in the option of an open fuel cycle. There are channels to dispose-of all the radwaste except the LL-LLW and IL-LLW and HLW (See Appendix 3). For the former, Andra geological investigations are in progress to site a sub-surface repository in clay. For the latter, Andra plans to ask ASN to license Cigeo by the year of 2019. Cigeo will be a deep repository sited in a clay layer at 500 meters depth and 130 meters thick designed to accept all the radwaste that cannot be disposed-of in sub-surface repositories. Radwaste to be considered is that produced and to be produced by all the reactors and facilities of the present nuclear fleet and fuel cycle, whatever the future energetic choices that will be made by the government. The present choice is to recycle once plutonium and uranium from $UO_x$ spent fuel and to store MOX spent fuel as valuable nuclear matter for launching fast neutron reactors. It is assumed that all $UO_x$ spent fuel will be reprocessed.

The sorting of radioactive substances into radwaste and nuclear matter with added value and available for later use, is the responsibility of the producers.

In China, the system of radwaste categorization is developed on the basis of IAEA safety standards of radwaste classification and the current version is the adoption of the equivalent of the *Classification of Radioactive Waste* (GSG-1) issued by IAEA in 2009 (Ref. [29]). The classification of radwaste is similar to that of France, which is set up according to the disposal strategy. The major differences are that the low-level radwaste in China corresponds to low and intermediate short-lived waste plus low-level long-lived radioactive waste with lower specific activity in France, and the low-level long-lived radioactive waste with higher specific activity in France corresponds to the intermediate radioactive waste in China.

## 3.3 Processing and discharge of radwaste

### 3.3.1 Minimization of radwaste

The risk that the management of radwaste leads to environmental impacts is lowest when the amount of crude radwaste to be processed is the lowest. The minimization of the quantities of radwaste starts by sorting out the radioactive substances produced in all facilities. It allows eliminating those characterized by a radioactivity that is at the detection limit or under clearance level, if they exist. The next step is the packaging of radwaste to reduce the dispersion of radionuclides in transport operations and storage. There are many packaging techniques for finding the economic optimum between any immediate environmental impacts due to packaging and storage and delayed environmental impacts due to geological disposal. In all cases BATs are generally implemented.

Nearly all countries have clearance levels or detection limits of radwaste, which lead to the de-categorization of potential radwaste in non-radioactive material. Such releases of materials for public uses can result in a substantial reduction of the most abundant VLLW. They concern the concepts of exemption and of clearance of radioactive materials. The first concept relies on the definition of activity concentration ($Bq \cdot g^{-1}$ or $Bq \cdot cm^{-2}$ or total activity) for limited quantities of matter (1 ton for instance), below which no control is necessary to assure radiological protection or that environmental impacts are negligible when for instance recycled materials are used. The second concept relies also on the consideration of the activity concentration ($Bq \cdot g^{-1}$ or $Bq \cdot cm^{-2}$ or total activity less than, or equal to, those for exemption) for possible re-use of decontaminated materials. Universal clearance levels are such that for any pessimistic scenario the radiological impact is less than 0.01 mSv per year (recommended dose by IAEA-safety rule RS-G-1.7 and Euratom-directive 96/29). Such low doses would not have an impact on the environment.

The other way to minimize radwaste quantities is the recycling of LLW like metallic materials. They can be melted in such a way that the processes lead to their decontamination. Melting is the only process which leads to radioactivity homogenization of the material for recycling, later facilitating their monitoring.

It seems impossible to reduce the quantities of other radwaste produced along the fuel cycle by recycling.

France does not practice the release of VLLW. The French Nuclear Safety Authority (ASN) considers that every material which has possibly been in contact with a radioactive contamination or which has been activated by radiation is a VLLW subject to regulation. The main reason claimed by ASN for not applying exemptions and clearances is the difficulty of the application of these concepts so that the limit of 0.01 mSv/a is ensured for the added dose to an individual. ASN considers that, against the advantages of clearance, it is impossible to consider all possible scenarios; they point out that parameters of safety analysis are subject to discussion, protocols of measuring radioactivity are difficult to implement at industrial scale and finally that there is a risk to make artificial radioactivity as ubiquitous, as natural one. This position is not coherent with international recommendations and some practices in Europe, and it is presently being reviewed. Nevertheless, some exceptional authorizations of clearance could be given (conditional clearance) for special cases in which the addition of radionuclides to solid materials (except products in contact with human beings) could be monitored. It might also be possible to recycle in the nuclear industry for special materials only contaminated at very low level of radioactivity and that can be monitored. Cases are submitted to ASN for specific approval.

For several years, the waste producers, Andra, IRSN and ASN have been studying the conditions for creating a release threshold for VVLLW (very, very low-level waste). EDF and Orano are looking to the technological and economic conditions to recycle by melting large quantities of metallic radwaste.

The safetyguide on radioactive waste minimization is in place in China. The newly constructed NPPs comply with the requirements of this safety guide. NPPs under operation have taken practical measures to implement the principle of radioactive waste minimization.

### 3.3.2 Discharge of effluents

Gaseous or liquid releases to the environment are the main sources of immediate environmental impacts, as already indicated. In the case of waste management, it is the question of the effluents associated with the packaging of primary waste. Gaseous discharges are decontaminated by filtration and/or by washing with appropriate aqueous solutions, if necessary. This leads to solid secondary waste and decontaminated gases. These gases are released to the atmosphere according to regulatory requirements.

The liquid effluents originating from the processes implemented in nuclear facilities are treated locally to produce decontaminated liquid solutions which are released to the environment in compliance with the authorizations, which produces solid radwaste and concentrated radioactive liquid, which are converted to a solid form.

In China both the radioactive gaseous waste and radioactive liquid waste need to be properly treated to make their radioactive levels as low as reasonably achievable while meeting the discharging requirements. All emissions of gaseous and liquid effluents are monitored and controlled to ensure that no accidental releases occur. The liquid effluent is discharged, and the appropriate discharging point is properly selected to optimize dilution in receiving water, and meet the near zero emission requirement for inland site NPPs.

## 3.4 Disposal of radwaste

### 3.4.1 Very low-level waste (VLLW)

For radwaste with a low/very-low activity (less than $10^2$ Bq/g), even containing trace amounts of long-lived radionuclides (like uranium), the landfill disposal concept (on surface or sub-surface) is generally adopted by most nuclear countries around the world. In fact, as radioactivity is low, the half-life of radionuclides is not a determining factor. In general, there are large quantities of such radwaste. The difference with disposal of short-lived radwaste is that it requires only light packaging and a rather limited engineered infrastructure. Packages, if any, have no function of confinement of radionuclide or toxic materials. Regulations defined for short-lived radwaste also apply to this category: control of radwaste (packages, big-bags, materials, etc.), control of filling of facilities according to predefined capacities, control of releases, and control of the environment. There are many types of disposal facilities around the world for technological radwaste, radwaste originating from processing of yellow-cake only containing natural radionuclides as well as radwaste produced through uranium enrichment. IAEA recommend surface/subsurface management by trenching.

France will have to dispose-of around more than 2 billion cubic meters of VLLW which exceed by a factor 4 to 5 the capacity of the present sub-surface repository. The major part of VLLW will come from dismantling of reactors and nuclear facilities. Andra will extend the capacity of the present repository. Andra, radwaste producers, ASN and IRSN are looking for a new management approach of VLLW: addition of a new central disposal, possibly decentralized disposal centers, recycling of metallic radwaste and concrete, conditions for releasing VVLLW.

As indicated previously, France does not practice the release of VLLW.

There are four landfill facilities operated for VLLW disposal in China. Around 10000 $m^3$ of VLLW have been disposed of so far.

### 3.4.2 Low and intermediate level-short lived waste (LIL-SLW)

Short-lived radwaste ($10^2$ - $10^6$ Bq/g) mainly originates from NPP operations. Some contain very small amounts of long-lived radionuclides. Packages of short-lived waste are in general disposed-of in specifically engineered surface/subsurface facilities. The depth of sub-surface facilities could be of several tens of meters. Packages take the form of steel or concrete drums or large containers, sealed or not. Safety and environmental authorities define the radiological capacity, and the capacity for each radionuclide or toxic material, that can be accepted up to the closure of the repository. These capacities depend on the characteristics of the site, the structures designed to host the packages and the engineered barriers. The capacity limitations for surface disposal facilities consider the perspective that they could return to a greenfield status after several hundred years when short-lived radionuclides will have disappeared, but not the long-lived ones. The authorized releases of gas and liquid effluents are also regulated and controlled, and the environment is monitored. All authorizations are set in accordance with the safety analysis of the disposal site/facility.

During operation, which may cover several decades, the main vector of transfer of chemical species from packages to the environment is rainwater or subsurface water flows. Rainwater is collected and processed if necessary. Ground water is monitored at the outlets of the site if this is needed. There is also the possibility that some gases escape the packages, like tritium in the form of tritiated hydrogen and tritiated water and this is why tritium capacity of the repository is limited.

Siting, operating and monitoring of repositories for short-lived radwaste enables the testing of new technologies to improve the confinement of radionuclides, thus reducing immediate or long-term releases of radioactivity into the environment and becoming a reference for surface/subsurface radwaste management.

When the disposal site will be returned to greenfield status, impacts are expected to be limited to those evaluated in the safety case analysis.

Waste containing large amounts of tritium is kept in storage facilities waiting for a decrease of tritium activity and is then managed according to the industrial channel adapted to its classification.

France will have to dispose-of around 1.5-2 billion $m^3$ of LIL-SLW coming from the present nuclear fleet. The capacity exists. The return of experience from 25 years of operation of a closed LIL-SLW repository (0.5 million $m^3$, closed more than 20 years ago) indicates that tritium is difficult to confine but that the impact on the public is less than a fraction of one μSv/a (See Appendix 3).

The low-level radwaste mainly contains short-lived radionuclides and limited number of long-lived radionuclides in China. This type of waste can be disposed of in near surface disposal facilities, which corresponds to Low and intermediate short-lived waste in France. China has disposed of about 20000 $m^3$ of radwaste in two near-surface disposal facilities (Ref. [30]).

### 3.4.3 Low level-long lived waste (LL-LLW)

This radwaste (10 to $10^5$ Bq/g) cannot be accepted in repositories for LIL-SLW or LLW because it contains some radionuclides such as $^{36}Cl$ or $^{14}C$, which are difficult to confine by engineered or natural barriers and, in addition, are present in quantities that are too large to be deposited in deep geological repositories. If a sub-surface disposal is considered, the site has to be selected according to the requirement of confining these radionuclides for a very long time. Therefore, the depth of disposal must be sufficient in order to guarantee a well-functioning natural barrier of adequate thickness.

The total amount of LL-LLW expected from the present nuclear fleet is around 190000 $m^3$. Andra is continuing to characterize a potential site in clay according to two sub-surface disposal-of concepts. A preliminary concept should be ready in a few years from now. The absence of a repository leads to prolonged storage of radwaste and slows down dismantling.

There is no low-level long-lived waste in the classification system of radwaste in China. The waste (10 to $10^5$Bq/g) containing long-lived radionuclides at lower levels of activity concentration than the upper limit of low-level waste belongs to low-level radioactive waste and can be disposed-of in near surface disposal facilities. The waste containing long-lived radionuclide at higher levels of activity than the upper limit of low-level waste would be categorized to intermediate-level radioactive waste and be applicable to intermediate-depth disposal.

### 3. 4. 4 Intermediate level-long lived waste (IL-LLW) and high level waste (HLW)

According to nuclear experts, the isolation of IL-LLW ($10^6$ to $10^9$ Bq/g) and HLW ($10^9$ Bq/g and more) from the environment and confinement of radionuclides can be assured in deep geological formations combined with multiple engineered barriers. Such formations must have been stable for hundreds of millions of years and feature favorable geochemical properties like limitation of water circulation and retention of chemical elements. The basic reason for choosing geological disposal for high level waste comes from sociological considerations about the stability of society that cannot be assured for more than a few centuries. It is then more rational to entrust geology in keeping this waste away from the biosphere for a very long period of the order of geological times.

Whatever the nuclear fuel cycle option chosen, after a long interim storage period (for example in cooling ponds or in dry storage) allowing a decrease of their thermal radiation, packages of HLW and IL-LLW will be disposed-of in deep geologic formations. To this purpose, special surface facilities are designed to accommodate the reception of such packages until their further handling. Primary packages are then placed in an over-package before being disposed-of. Environmental impacts during storage as well as during the dropping of packages into the repositories are the same as those encountered during normal or incidental operation of nuclear facilities, particularly with regard to authorized releases.

As already indicated a deep repository is designed to accept all the radwaste that cannot be disposed of in surface/subsurface repositories. There is no limitation of its capacity with respect to activity of radionuclides.

There are several concepts for deep repositories depending on the geological rock formation chosen for siting them, for instance clay or granite. Clay slows down and finally stops the migration of all radionuclides present in spent fuel because of its high capacity to catch them by various mechanisms. This is why clay is used as a buffer if the selected rock is granite. Extensive and detailed investigations have been carried out and are still underway in countries that need to find a site for their deep repositories.

Up to now only Finland has drilledshafts in granite to establish an underground spent nuclear fuel repository (Onkalo) down to about 450 m. Sweden is close to do the same. Spent fuel will be encapsulated in copper canisters and deposited surrounded by bentonite rings in wells drilled in the granite of the Scandinavian shield (KBS3 concept). All galleries and shafts will be filled with bentonite before the repository is sealed off. France is ready to apply for the license to install a repository in clay at 500 m in a few years from now for IL-LLW and HLW produced by reprocessing. In France over-packed nuclear glasses will be deposited in horizontal tunnels and over-packages of IL-LLW will be deposited in large cavities both excavated in the (vertical) center of an extended horizontal clay layer 130 m thick (Callovo-Oxfordian clay). All engineered structures as well as galleries and shafts will be sealed with special concrete/bentonite plugs. Such engineered structures are expected to confine radionuclides (and toxics) over a very long period of time (tens to hundreds of centuries) that will prevent any impact on the biosphere.

Ten other nuclear countries are more or less in an active preparation stage for several decades. They hope to open a repository during the next decades. Implementation of geological repositories spans long periods of time owing to the extensive processes of site characterization, analysis and final selection, involving large scale scientific studies as well as political and public participation in the decision-making process. All countries have produced numerous reports on their national programs to site an underground repository. International organizations (EU, OCDE-NEA, IAEA), have set up joint international research projects in that direction. These programs aim at the understanding of basic phenomena controlling migration of radionuclides and at testing engineered barriers.

It is expected that operation of a deep geological repository will last more than one and half century, as planned for instance in France. During this period of time, environmental impacts may be due to an accidental situation despite the measures taken to prevent them. The impacts on the environment can be anticipated by simulation and compared to conditions prevailing at present. The environment is also monitored during a long period of time prior to the opening of a repository.

The simulation of the long-term evolution of the components of a repository after its closure is the main issue of the safety case analysis. Despite their capacity to isolate and confine radionuclides, packages of IL-LLW and HLW will progressively be corroded and release of radioactivity will subsequently occur. The migration of radionuclides and other elements will then slowly start by diffusion. According to numerous leaching experiments and detailed examination of natural analogues, the life of nuclear glass or uranium oxide packages is estimated to be over hundreds of thousands of years. Results of numerous simulations demonstrate that migration of actinides could not be more than ten meters in clay and the time of mobile fission products to reach the biosphere would be so long that their activity would be drastically decreased.

Simulations of the migration of radionuclides into the environment enable the calculation of concentrations of long-lived radionuclides at the outlets of the site. Then, according to scenarios of land and water use, doses to people can be derived according to standard methods in use today.

Several simulations for periods of up to a million years or more showed that, owing to the efficiency of radwaste packages, of natural and engineered barriers in deep repositories, such release of radionuclides will lead to doses at ground surface that will not exceed one tenth of a percent of the exposure to natural background radioactivity (See Section 3. 1. 2).

The file presented to the nuclear safety and environmental authorities to obtain a license to open a geological repository contains all the data, results of experiments and of simulations. It includes the safety case evaluation of the repository.

According to the present fuel cycle strategy it is expected that around 72000 $m^3$ of IL-LLW and 12000 $m^3$ of HLW will need to be placed in the repository. This waste is in storage pending for the commissioning of Cigeo.

As specified in the current radwaste classification system in China, the intermediate level waste is defined as waste that contains long-lived radionuclides in quantities that need a greater degree of containment and isolation from the biosphere than is provided by near surface disposal. Disposal in a facility at a depth of between a few tens and a few hundreds of meters is considered for ILW. Disposal at such depths has the potential of providing a long period of isolation from the accessible environment if both the natural barriers and the engineered barriers of the disposal system are properly selected. In particular, there is generally no detrimental effect of leaching at such depths in the short to medium term. Another important advantage of disposal at intermediate depth is that, in comparison to near surface disposal facilities suitable for LLW, the likelihood of inadvertent human intrusion is greatly reduced. Consequently, long term safety for disposal facilities at such intermediate depths will not depend on the application of institutional controls.

A geological repository is planned to be constructed around years of 2050 (Ref. [31]). At present, the general program of R&D of geological disposal was completed in China. The Beishan area in Gansu Province has been determined as the primary pre-selected site region for geological disposal of high-level radwaste. The material of buffer and backfill for geological disposal has been developed and research on radionuclides migration and safety assessment is in progress. The site and construction programme of the Beishan underground research laboratory (URL), which is the first URL of China, have been determined. The Beishan URL is currently under construction and it is expected to be completed in 2027.

### 3.4.5 Radioactive waste containing only natural radionuclides from the front-end of uranium fuel cycle (uranium mining)

Large quantities of uranium radwaste from uranium mining consist of tailings and waste residues from ore processing (to get the yellow-cake) and additional technological waste. This radwaste contains uranium and all its non-volatile daughters as well as other chemicals ($^{226}$Ra is the only one present in a sizable amount).

The refining of yellow cake and its transformation to gaseous fluoride to be enriched in $^{235}$U yields large quantities of radwaste containing only natural radionuclides.

### 1. Mining

In France, uranium mining has been operational during a period of 50 years at 250 sites producing 80000 tons of uranium from 52 million tons of ores, and is now discontinued. Mining radwaste accounts for 166 million tons of excavated rocks including tailings and 52 million tons of mining residues. The orders of magnitude of uranium content and radioactivity levels usually associated with materials and residues on mining sites are shown in Table 3 - 1 (Ref. [20]).

**Table 3 - 1 Uranium content and radiation in uranium mining activities and natural rock formations**

| | Uranium content/(g/t) | $^{226}$Ra activity concentration/(Bq/kg) | Overall activity concentration/(Bq/kg) |
|---|---|---|---|
| Average of soils and rocks in France | A few | A few tens | A few hundreds |
| Granitic rocks | A few tens | A few hundreds | A few thousands |
| Ore | About thousand | A few tens of thousands | A few tens of thousands |
| Steriles | A few tens to hundreds | A few hundreds to a few thousands | A few thousands to a few tens of thousands |
| Residues | A few hundreds | A few tens of thousands | A few hundreds of thousands |

Mining waste is disposed-of *in-situ* in large excavations and at closure covered with a layer of natural material to prevent radon emanation and direct gamma exposure. Monitoring concerns radon emanations, uranium and toxic releases in rainwater and underground water at the facilities' outlets. Water is processed and after decontamination released to the environment, only featuring traces of uranium and radium. Accumulation of these radionuclides in the environment is monitored and periodical remediation follow if necessary. Considering the various uranium fuel cycle steps, mining, milling and leaching the ores are the most polluting steps.

In France, residues and tails are stored in 17 ICPE-classified repositories. The residues (clay sands or blocks of ores leached by $H_2SO_4$) are placed on a geo-polymer basement and are under cover (2 m of steriles, 0.4 m of soil). The percolation water is treated (6000 $m^3/a$) either before discharge or to recover part of the uranium. Feedback from monitoring and analysis of core samples in residues indicates a certain stabilization of residues after alteration by water and diagenesis. In addition, U and Ra are trapped by some mineralogical phases. The uranium is adsorbed on clay minerals and oxy-hydroxides of Fe (Ⅲ) and also forms insoluble U (Ⅵ) /U (Ⅳ) mixed phosphates. The radium co-precipitates with $BaSO_4$ and is also adsorbed on the clay minerals. Modeling of the behavior of U and Ra has been achieved. In total, U and Ra are not very mobile. All the information gained from environmental monitoring and in-situ periodical analysis may be used to model the migration of these elements in the long-term.

The French experience at home (and in Niger) indicates that the main problem still requiring attention is the possible re-use of tailings and mining waste as rocks in construction or ballast materials (and in the case of Niger also the re-use of contaminated scrap iron). In France restoration of radiological standards with regard to the dispersion of sterile tailings is applied when the local impact is greater than 0.6 mSv/a. All tailings and mining waste rock are now in repositories. The impact is most significant for buildings erected on tailings and mining waste rock as the induced individual doses are in the range 0.5 to 1 mSv/a and the radon concentrations yield 1000 $Bq/m^3$. In France health regulations have laid down the limit for the public at 1 mSv/a and that 300 $Bq/m^3$ of radon is the limiting value for this dose rate. Frequenting areas ballasted with tailings still in place, induces significantly lower doses (at least one order of magnitude) and is consequently not a problem.

Eighty sites for uranium mining have been constructed in China and about thirty of them have been decommissioned. Mining radwaste accounts for 34 million tons of excavated rocks, including tailings and 11 million tons of mining residues (Ref. [32]).

Orano operates uranium enrichment in France. The radwaste is managed on the site.

**2. Radwaste from refining Yellow cake**

Orano conversion facilities are sited at Malvesiin southern France. The radwaste produced up to now have been left on site, the fresh aqueous nitrate effluents in large ponds (70000 $m^3$) and the others are stored (around 280000 $m^3$) in sub-surface deposit. The solid nitrates collected from nitrates effluents in "evaporation ponds" will be processed to become VLLW and IL-LLW, included in the present inventories. A new facility has recently been commissioned. The future radwaste coming from a new process will be managed in line as VLLW and LL-LLW.

## 3.5 Open/closed nuclear fuel cycle

Environmental impacts due to waste management are linked to the radionuclides released from reactors and facility operations (including mining) and to the quantities of radwaste produced. These indicators enable comparisons between nuclear fuel cycles. The estimates from the French CEA (Ref. [33]) for open fuel cycle (OFC) and closed fuel cycle with single-recycling of Pu (CFC) actually operated in France are shown in § 2.4.1. This section draws attention to the fact that reprocessing in CFC releases noticeable quantities of noble radioactive gases and tritium ($5.5\times10^{11}$ Bq/TWh) into the atmosphere as well as some slightly radioactive liquids to the sea ($2.24\times10^{10}$ Bq/TWh) but without significant radiological impact. For the two fuel cycles, the production of LLW and LIL-SLW is not significantly different, however, CFC produces 4 times more IL-LLW (1.18 $m^3$/TWh versus 0.32 $m^3$/TWh) than OFC and it is the reverse for HLW (0.36 $m^3$/TWh versus 1.17 $m^3$/TWh).

The local environmental impacts, due to the operation of repositories, are facility specific. Operators report annually to safety and environmental authorities. Regarding radiological impacts, public doses are estimated to be less than 1/10 μSv. The objective for long-term impacts is to comply with the present radiological target of less than 1 mSv/a in all cases.

Taking into account, for a CFC, that the present reactors are replaced by Gen-Ⅲ reactors (typically EPR), the production of radwaste will be reduced by 20%~35%, depending on the types; these results are due to the better performances of Gen-Ⅲ reactors in terms of thermodynamic efficiency, higher burn-up and in-service life. Nevertheless, liquid releases increase by about 20% mainly due to reprocessing and reactor operation, but their radiological consequences will remain low compared to natural radiation.

## 3.6 New technologies

If electricity was produced by SFR, or more generally Gen-Ⅳ fast neutrons reactors, this would lead to a drastic reduction of releases and waste production due to the elimination of all the operations of the front-end cycle. Another significant improvement in the reduction of long-lived radwaste impacts would be the extraction of minor actinides (Am, Cm, Np) from the spent fuel during the reprocessing, with the use of Gen-Ⅳ fast reactors or hybrid reactors for burning them. Theoretically this additional step in spent fuel reprocessing would permit to shorten the length of time for hazards from HLW from several hundreds of thousands of years to only hundreds of years. In France, the demonstration of the feasibility of such an extraction of minor actinides has been demonstrated at the pilot level on a kilogram scale. The extension to the industrial scale would become possible if and when Gen-Ⅳ fast reactors will come to maturity. It has been shown that transmutation of minor actinides produced by an SFR fleet is only possible if all the SFRs of the fleet are able to transmute actinides. That requirement implies a drastic change in nuclear electricity production: new SFRs, new closed fuel cycle, new extraction methods, new fuel fabrication, etc. Regarding the environment, the more radioactive matter is submitted to chemical processes the greater is the risk of radionuclides release.

In addition, it is obvious that for economic reasons, transmutation of minor actinides could not be applied to material that has already been packaged in nuclear glasses.

Another way of transmutation of minor actinides is under investigation using an accelerator driven system (ADS). The most advanced project is Myrrha in Belgium. There is only prospective information about the impacts, if any, of ADS-transmutation on environment. Whatever the performance of ADS might be, it will be necessary to prepare the transmutation targets and probably recycle them to get a good transmutation yield. Separation of radioactive material always leads to a limited, but unavoidable, release of radionuclides to the environment.

During the last ten years, France has developed an ambitious research programme to be ready to launch a first commercial SFR in the frame of Gen-Ⅳ aroundthe year of 2040 but this target is now reconsidered.

The ambition of China is to have a first commercial SFR aroundthe year of 2035 and to deploy large-scale construction around 2050.

A project on accelerator driven system (ADS) for the partition and transmutation has been initiated in China and the demonstration project is planned to be constructed around 2050.

## 3.7 Conclusions

Environmental protection constraints are basically taken into account at each step of radwaste management-isolation/confinement in packages, storage, and disposal in facilities adapted to each type of radwaste. All measures benefit from top level technological developments and are supported by continuous R&D on the behavior of radionuclides/toxics in engineered barriers and the geosphere.

Monitoring is carried out during all operations from production of radwaste up to its disposal in repositories which isolate radwaste packages from the biosphere. The background level is permanently monitored around these facilities. Normal releases do not have an impact on the environment greater than that authorized by safety and environmental authorities when the facilities were licensed. Feedback from experience shows that such releases are far smaller than initially expected. Unusual releases are promptly detected.

The main environmental impacts are induced by the front-end of the nuclear fuel cycle.

The assessment of radiological and chemical impacts on people is the responsibility of radiation protection and health protection authorities. They are based on reliable scientific data and tested models of irradiation and incorporation of radionuclides. However, the R&D continues to reduce uncertainties on the data and to improve the models and this effort needs to be maintained.

Estimates of ionizing radiation impacts on ecosystems are less well supported, and R&D effort on this subject needs to be increased.

After closure of the repositories, monitoring will continue during a test period; then safety will change from active to passive. Most radionuclides will decay in the repositories, those that might return to the biosphere will do so at a time so long that their radiotoxic impact will be negligible.

As of today, the WIPP repository is in operation in USA, New Mexico, which has been designed in a deep salt formation, to accommodate transuranic radioactive waste left from the research and production of nuclear weapons. No deep geological repository accommodating HLW from commercial nuclear power is in operation yet. In 2015, the Finnish Government issued the construction licence of the disposal facility in ONKALO, which is the first construction license of repository in the world. Besides, the HLW disposal projects in Sweden and France are presently at licensing stage.

In terms of society it is important that the proven or potential impacts of radioactive waste management on environment be brought to the public's attention in a transparent manner.

# Appendix 3

## A3.1 French side

It is instructive to consider the feedback of measures taken in France with respect to the environmental impact of radwaste management. This information originates from the monitoring of three types of repositories: the first one is now closed, the second one is currently in operation and the third one is to be constructed.

### 1. CSM

The surface repository for LIL-SLW called CSM (Centre de Stockage de la Manche) was in operation for twenty-five years (1969 - 1994) and the construction of a provisional multi-barrier cap lasted six years. The repository is sited near the sea on a basement of gneiss rock. A total of 1470000 packages (530000 $m^3$ of waste) have been deposited over 15 hectares. Several civil engineered structures to host packages have been experimented. Since the year 2000, CSM is under monitoring (tests of confinement of the cap and structures, tests on surface and underground water yield relying on 10000 measurements per year). This facility will be closed after completion of a definitive cap, expected to take place around the year of 2050 and 2060. Monitoring will continue up to the time when passive safety is reached. Feedback of CSM monitoring is that tritium is difficult to confine. Maximum radiological impact on the environment is 0.20 μSv/year for the full use of the water stream flowing in the near vicinity of this repository.

### 2. CSA

The CSA (Centre de Stockage de l'Aube) surface repository for LIL-SLW is in operation since 1992 with a surface area of 95 hectares. It is sited near Soulaines-Dhuys (Aube department) on layers of sand and clay. This center is designed for the disposition of $10^6$ $m^3$ of radwaste. Exposition to radiation of the most exposed group of people at and around the site shall not exceed 0.25 μSv/a. Management of CSA benefits from the feedback from CSM (See above). Standardized packages are placed in engineered structures sheltered from rainwater. When structures are full, packages are drowned in concrete to form a monolith. CSA is monitored in terms of radiological, physico-chemical and ecological parameters (15000 checks per year). According to the measured release of several radionuclides to the environment, the radiological impact is around $1 \times 10^{-3}$ μSv per year. Tritium is not detected neither on nor around the site, or in the underground water.

**3. Cigeo**

In the coming years, France will open the deep geological repository Cigeo for IL-LLW and HLW near Bure (Meuse/Haute Marne departments). Packages will be deposited in engineered structures constructed in a clay layer 130 m thick at 500 m below the surface. Surface facilities will receive and condition packages to be transported underground. Operation of Cigeo is expected to last 150 years. The environment of Cigeo will be monitored; Andra operates since 2007 an environmental permanent observatory (EPO) extending over 900 $km^2$ with a reference sector of 250 $km^2$ around the Cigeo facilities (with a meshing of 1.5 km×1.5 km). The objective is ①to understand all the impacts on air, water, soils, flora and fauna through measurement of the physico-chemical exchanges between them with respect to human activity and ②to record and retrieve these data. The EPO is associated with a facility to preserve samples. There are 2500 observation points, collecting 2500 samples and 85000 data samples per year.

## A3.2 Chinese side

### Northwest disposal site

The Northwest disposal site is located in Gansu Province, north western China. It receives, stores and disposes of low level and intermediate level radioactive solid waste for near-surface disposal. The construction of the northwest disposal site began in 1995 and was completed in 1998. In 2011, it was approved for operation. The planned disposal capacity for this Site is 200000 $m^3$. The disposal capacity for the first phase is 60000 $m^3$, consisting of seventeen disposal units. To date, six units have been built, the capacity of which is 20000 $m^3$. The site is located in the cohesive soil and sandy soil interbed with a thickness of approximately 50 meters. The concept of the disposal unit is a reinforced cement structure. Sandy soil is back filled between waste drums and between waste drums and disposal unit wall. The disposal unit will be poured with reinforced cement to form top plate when it is full. After closure, the top of each disposal unit will be finally covered with a two-meter-thick cap. During the disposal facility construction, reinforced bottom plate (slab) was added for higher safety. The safety-assessment for the unintentional intrusion scenarios during institutional control after closure includes the dwelling, drilling and water well-digging on the disposal site. Results of the analysis indicate that the exposure dose received by the unintentional intruder for these scenarios is not more than 0.1 mSv/a, which is much lower than the national limit of 1 mSv on annual effective dose to the general public. Northwest disposal site had received a total of about 22000 $m^3$ solid radwaste by 31 Dec., 2019.

# Chapter 4　Severe nuclear accidents

Recommendations

- It is necessary to continue research and development on the mechanisms leading to severe accidents and provide support for their mitigation. It is suggested that further studies on measures of maintaining the integrity of containment and development and application of advanced technology (such as ATF) should be carried out.
- It is necessary to further accumulate experience in the implementation of severe accident management guidelines and to pursue the development of mitigation measures aimed at coping with large-scale damage in NPPs, multi-units' accidents and to strengthen emergency response capacities.

## Introduction

Since the peaceful use of nuclear power in the 1950s, after years of development, nuclear power, coal-fired power and hydro-power have been known as the three main sources of electricity. The integral safety situation of the world 454 nuclear units is good (according to data from IAEA PRIS), and more than fifty years of normal operation of commercial nuclear reactors proves that the radiation impact of nuclear reactors is extremely low and much lower than the natural background radiation level (cf. the joint report of the three Academies about the future of nuclear power). However, the accidents at Three Mile Island (TMI), Chernobyl, and Fukushima Daiichi NPPs accidents have had a major impact on the development of nuclear energy and on the world view of nuclear generation of electricity. It is at this stage important to review the above nuclear accidents, and to summarize the design improvements and measures taken by the nuclear industry to reduce the severe accident frequency and limit any post-accident consequences, and consider feedback of experience.

This chapter begins with a review of severe accidents and an account of their impacts as well as of lessons learned from them (Section 4.1). The measures taken to avoid such events through a continuous improvement in technology and management are described in Section 4.2. Conclusions are given in Section 4.3. In addition, in order to further clarify the technical and policy consideration for severe accidents in China, three specific items are included at the end of this chapter. Appendix 4-1 introduces dedicated prevention and mitigation measures for severe accidents of Gen-Ⅲ NPPs. Appendix 4-2 introduces the safety issues of inland NPP sites, and Appendix 4-3 introduces emergency management after severe accidents in China.

## 4.1 Severe accidents

Three accidents have had a notable impact on nuclear industry worldwide. These accidents took place at the Three Mile Island NPP in the United States in 1979; at the Chernobyl nuclear power plant in the former Soviet Union in 1986; and at the Fukushima Daiichi nuclear power plant in Japan in 2011. The former was ranked at level 5 on the IAEA INES (International Nuclear Event Scale System) with no or limited consequences on the environment. The latter two accidents, qualified as severe, have been rated at level 7 based on the IAEA INES; they have had serious consequences on the environment.

The causes of these accidents, their impacts on the environment and the lessons learned differ in most respects, but these severe accidents have substantially promoted progress in nuclear safety technology and augmented nuclear safety level. The new safety features and dispositions are aimed at reducing the environmental impact even in case where such an accident might happen again, and this may hopefully diminish people's worries. Today, the designs of new reactors around the world have been significantly improved giving rise to Gen-Ⅲ NPPs. Moreover, with the accumulation of operating experience, the management ability of nuclear units has been effectively improved, so that even in the worst situation, the risk of radioactive material release to the environment is decreased to a very low level. In parallel safety authorities have published guidelines for emergency in case of accident as well as for remediation. Regulations have also moved to reinforce the obligation of operators to apply these guidelines.

This section analyses the three accidents and the changes that they have induced on the technical and management levels and in terms of regulations. Each accident is documented with a focus on information related to improvements of the impacts of nuclear energy on the environment.

### 4.1.1 Three Mile Island accident (Ref. [34])

**1. Cause of accident**

The TMI NPP employed pressurized water reactors developed at an early stage in the United States. The cause of the accident was equipment failure, inadequate interpretation of the state of the system by the operators and subsequent inappropriate decisions. As a result, the reactor core melted, and a large amount of fission products entered into the containment. Fortunately, the containment maintained its integrity and confined the major part of radioactive substances produced in the accident.

**2. Impact on the environment**

The accident produced a limited release to the environment. The maximum dose to the surrounding public was ten times less than the doses from the annual natural background. No casualties were caused and there was no mid-or long-term impact on the environment.

### 3. Lessons learned

Although the Three Mile Island accident only brought minor radiological consequences to the environment and the public, and caused no casualties, the direct economic losses were huge. It sounded the alarm for the entire United States nuclear industry and regulatory authorities and has had far-reaching implications for the development of the world's nuclear industry. The TMI accident also had a notable impact on the nuclear industry that was in rapid development at the time in the US, Europe and in other countries.

The TMI accident proved the overall soundness of the safety concept based on defence in depth, but also revealed weaknesses and deficiencies in design, management and safety studies. It showed that small details that had not been considered before were capable of producing serious consequences. The accident indicated that management aspects (e. g. , operator training, emergency procedures, organization and coordination) had no lesser importance than technological aspects (e. g. , equipment design, construction, qualification, and safety analysis).

After the TMI accident the nuclear industry made substantial improvements in man-machine interactions, monitoring, control and training of plant operators. A significant move took place in the safety analysis under accidental conditions where nuclear companies and regulatory agencies acted in concert to shift their focus from reactor research targeting DBA (design basis accident) to research on severe reactor accidents, and initiated large scale research projects on severe accidents. As a watershed, safety analysis turned from studies on LBLOCA (large break loss of coolant accident) to studies on SBLOCA (small break loss of coolant accident) and transients. The WASH-1400 report released in 1975 by a research group led by Professor Rasmussen adopted a probabilistic risk assessment (PRA) methodology demonstrating scientific capabilities surpassing those of more traditional deterministic analysis techniques. One outcome of the TMI accident was a strengthened interest for PRA. It established their viability and merits during decades that followed.

### 4.1.2 Chernobyl accident

**1. Cause of accident** (Ref. [35])

One of the causes of the Chernobyl NPP accident was linked with the stability characteristics of the RBMK reactors. The reactor was a graphite-moderated water-cooled core featuring a positive void coefficient with the potential risk of prompt super-criticality. There was no-containment vessel to confine radioactive substances in an accidental situation. Erroneous interpretation of the reactor state led to inadequate actions by the operating team resulting in a prompt super-criticality event. The sharp power increase led in turn to the explosion of the reactor. Large radioactive leaks occurred, which resulted in a sizable radioactive release into the atmosphere. The Chernobyl accident was primarily due to flaws in the management of NPP operation and to an insufficient nuclear safety culture.

The Chernobyl reactor was one of the seventeen RBMK reactors that were designed and constructed by the former USSR; it has been deployed in the Soviet Union only, and the concept was abandoned after the accident.

**2. Impact on the environment**

Major releases from unit 4 of the Chernobyl nuclear power plant continued for ten days, and large areas of Europe were affected to some degree by the Chernobyl releases. Much of the release comprised radionuclides with short physical half-lives; long lived radionuclides were released in smaller amounts (Ref. [36]).

One hundred and thirty-four emergency workers suffered an acute radiation syndrome, of which 28 died from radiation. Among the recovery operation workers exposed with moderate doses, there are some evidences of a detectable increase in the risk of leukemia and cataract. The occurrence of thyroid cancer among those exposed during childhood or adolescence has significantly increased due to the drinking of milk contaminated with radioactive iodine during the early stage of the accident (Ref.

[37], [38] and [39]).

Construction of the shelter was aimed at environmental containment of the damaged reactor, reduction of radiation levels on the site and the prevention of further release of radionuclides off the site (Ref. [36]). The radioactivity is continuously monitored, and periodic international reviews assess the evolution of the situation. Natural life is re-developing, and studies are now in progress to assess whether there are genetic effects on plants and wild animals.

### 3. Lessons learned

The Chernobyl accident and its post processing have set a great financial burden to the former Soviet Union states, Ukraine, Belarus and Russia, and have had a huge impact on local population and activities, and on the nuclear industry worldwide. The accident raised issues of public safety and public concern about safety, affecting the planning of the nuclear energy infrastructure. The depth and scope of this accident had an impact that was far beyond that of the TMI accident.

Lessons learned within nuclear industry have been again quite substantial:

(1) After the accident, the nuclear industry essentially abandoned core design concepts featuring positive feedback characteristics, ending graphite moderated reactors development. Inherent reactor safety features were augmented.

(2) Reactor protection systems were improved, and operators in main control rooms were subjected to more restrictions to effectively reduce the possibility of improper operations associated with human errors.

(3) Containment buildings as the last safety barrier were adopted by the whole industry further reducing the possibility of large radioactive releases to safeguard public health and environmental safety.

(4) Safety culture came into being and received greater attention from the nuclear industry across the world. Awareness of nuclear safety was extended widely, from NPP operation to design, manufacturing, construction, supervision and control, playing an important role in preventing nuclear accidents.

(5) Ideological isolation in nuclear technology that flourished under the context

of the Cold War was essentially waved out. The IAEA developed and implemented the *Nuclear Safety Convention* which calls for International peer reviews of regulators. International organizations such as WANO were established, to encourage operators to improve operational safety and foster the concept of "Borderless Nuclear Safety".

Although the consequences of the Chernobyl accident were quite serious and far-reaching, follow-up studies of the accident indicated that nuclear safety is guaranteed as long as safety guidelines are observed, safety awareness is enhanced, and safety design is constantly optimized. When all these items are considered, they ensure that newNPPs have a higher level of safety and that nuclear power remains a safe source of energy.

### 4.1.3 Fukushima Daiichi accident (Ref. [40] and [41])

#### 1. Cause of accident

Fukushima in Japan is located near the "sub-duction zone" of the Eurasian plate and the Pacific plate, which witnessed frequent earthquakes in the past geological history. The Fukushima Daiichi NPP adopted a boiling water reactor type of the earliest commercial reactor technology developed in the United States. Its design and construction were completed prior to the Three-mile Island nuclear accident when serious accidents had not been experienced and complex accidental sequences were not foreseen.

The triggering event of the Fukushima Daiichi accident was a super earthquake and subsequent tsunami with an amplitude that by far exceeded the design standard. Severe damage caused by the magnitude 9 earthquake and subsequent tsunami to infrastructure such as transportation and power systems in the surrounding areas deferred recovery of offsite power for 9 days after the earthquake, a time period that far exceeded design considerations. The four reactors damaged by the accident (out of the six reactors of the Fukushima Daiichi NPP) were located near the sea shore, to minimize the length of the cooling circuit. In addition to the increased exposition to tsunamis, this also had the effect of exposing diesel generators required in case of a loss of offsite power. The lack of tightness of diesel generator rooms, and the flooding of their air intakes ended up with a station black out and a disastrous loss of core cooling. It has also been pointed out that the lack of passive hydrogen recombiners led to explosions which sent radioactive materials in the atmosphere. In addition to the reactors the loss of cooling of spent fuel pools posed a serious threat of additional emission of radioactive materials to the atmosphere.

Finally, failure to evacuate residual heat from three operating units and spent fuel storage pools caused core melting, hydrogen generation and accumulation leading to explosion, and release of radioactive material in the environment. Lack of prevention and mitigation measures in case of severe accidents in NPP design had serious consequences.

### 2. Impact on the environment

Fukushima Daiichi accident resulted in the release of radioactive gases (less than 500 PBq of radioactive iodine, less than 20 PBq of radioactive cesium) and materials into the environment. Although dose rates exceeded some reference values in the early phases of the accident, no impact on animal and plant populations and ecosystems is expected. Long term effects are also not foreseen as the estimated short-term doses were generally well below levels at which highly detrimental acute effects might be expected and dose rates declined relatively rapidly after the accident.

People within a radius of 20 km from the site and in other designated areas were evacuated, and those within a radius of 20-30 km were instructed to shelter before later being advised to voluntarily evacuate. Evacuation resulted in the loss of farms and businesses. The Japanese government has undertaken heavy restoration works to clean the area so that the population could progressively re-occupy their land.

### 3. Lessons learned

The accident was also caused by a lack of prevention and mitigation measures for severe accidents. As indicated, the Fukushima reactors were designed and built and completed prior to the TMI accident when there were no clear consideration of severe accidents and complex accident sequences. In these early reactor systems, there were design deficiencies of preventive and mitigating safety features.

Experience and lessons learned from the accident have served as references to improve the design of new NPPs and to enhance the operational management of operating NPPs. Feedback from the Fukushima nuclear accident has concerned a number of items:

(1) External events beyond design basis require more attention in NPP design and operation. Evaluations of natural disasters should be more conservative and for this it is important to consider scenarios of their occurrence, in a sequential or in a simultaneous mode and analyze their combined impacts on NPPs.

(2) The Fukushima accident has outlined the specific needs for absolute tightness of the emergency pumps.

(3) It also put a spotlight on the protection of spent fuel pools in terms of structure and requirement for permanent cooling.

(4) Safety of NPP needs to be assessed on a regular basis to incorporate knowledge updates, necessary corrective actions and compensation measures that are to be immediately implemented.

(5) It is necessary to ensure that instrumentation and control systems will maintain their functions in DBA allowing to monitor basic safety parameters of NPPs and to facilitate operation. Residual heat removal requires robust and reliable cooling systems that are capable of functioning both in DBA and BDBA (beyond design basis accident) conditions.

(6) Training, drills and exercises need to include hypothetical scenarios of severe accidents to ensure that operators are fully prepared and ready to take the best decisions.

After the accident, the IAEA established a direct link with the Nuclear and Industrial Safety Agency (NISA), Japan's official liaison for the accident through emergency arrangements, and shared information that was continuously updated and released to member states, relevant international organizations and the public in general.

Although it is still hard to clearly measure the extent of the damage and its impact on the global environment, the Fukushima accident, as the Chernobyl accident, induced a shock straining the world, raising concern about nuclear radiation and enhancing public worries about environmental disasters that could be caused by nuclear power. This led many countries to review the nuclear safety of their domestic NPPs.

After inspections and assessments, many countries confirmed their position to pursue the development of nuclear power and adopted measures to improve safety of existing installations and enhance their emergency response capacities. Although the Fukushima accident slowed down the development of the world's nuclear energy, it has also promoted progress and improvement of nuclear safety and management.

## 4.2 Improvements to make nuclear energy free of environmental impacts in case of accident

To summarize, we can say that after the TMI accident there has been a series of major improvements in equipment reliability, operator training, and man-machine interface for pressurized water reactor (PWR) NPPs in the world. After the Chernobyl accident, several countries have undertaken extensive research to improve the safety of NPPs, and to develop advanced nuclear power technologies on this basis. After the Fukushima Daiichi accident, various countries organized nuclear power safety inspections and adopted further measures to prevent or mitigate nuclear accidents.

### 4.2.1 Improvements in reactor technologies

Taking into account the lessons learned from the above accidents, the nuclear industry has implemented many important technical improvements to Gen-Ⅱ PWR NPPs for those under operation and those under construction. At the same time, the concept of

Gen-Ⅲ PWR nuclear power plant was forward based on the requirements of improving safety, availability and reliability of NPPs, in order to practically eliminate large radioactive release after severe accidents.

The concept of practical elimination was first proposed by Europe and was later adopted by the International Atomic Energy Agency (IAEA) and agreed upon by China's nuclear industry. According to this concept, if some conditions are physically impossible or extremely unlikely to occur with high confidence, the occurrence possibility of such conditions can be considered to be practically eliminated. New NPPs built in China will strive to achieve the possibility of practical eliminate large radioactive release in design, a goal that is clearly defined in relevant planning documents on prevention and control of nuclear safety and radioactive pollution.

The Gen-Ⅲ PWR has adopted the concept of defence in depth, that multiplicity, diversity, and physical isolation design principles are included to improve accident response in case of an accident and mitigation capabilities, to practicality eliminate large radioactive release after severe accidents. The particular goals of preventing and mitigating severe accidents include:

(1) Preventing meltdown;

(2) Maintaining the integrity of the reactor pressure vessel (RPV);

(3) Maintaining the integrity of the containment;

(4) Preventing radioactive release of spent fuel.

The Gen-Ⅲ PWR are equipped with advanced large containment buildings capable of enduring external natural disasters such as earthquakes, tornadoes, and the destructions due to human induced accidents such as fires and explosions, as well as intentional or accidental crash of a large commercial aircraft or other terrorist acts. They are able to withstand environmental conditions such as high internal temperatures; high pressures and highradiation level caused by a severe accident, and maintain integrity, avoiding the release of radioactive substances to the environment. Appendix 4-1 explains the countermeasures and special improvements introduced by the Chinese nuclear industry to prevent massive release of radioactive substances.

Nuclear power industries in France and China have both developed their own Gen-Ⅲ PWR technology, which are materialized in the EPR and the HPR1000. A first EPR is already connected to the grid, while the first HPR1000 project progresses well.

In addition, nuclear industry in China has made specific technical improvements to the safety of "inland NPPs" which is of particular concern to the public, such as "Near Zero Emission" of radioactive liquid effluent under normal operation and treatment of radioactive liquid waste under severe accident conditions; and their safety level meets the highest safety requirements implemented worldwide. See Appendix 4 - 2 for more details.

In order to improve the safety of NPPs, especially for PWR, nuclear industry has actively developed new technologies, like the new generation of accident tolerant fuel (ATF), to minimize the possible hydrogen production under accidental conditions, and to eliminate the possibility of hydrogen explosion. The technique of in-vessel retention after severe accident has also been considered by nuclear industry and many research institutes have carried out relevant work on enhancing heat transfer and increasing critical heat flux (with for example an application of nanofluids).

### 4.2.2 NPP action after Fukushima Daiichi accident

After Fukushima Daiichi accident, self-inspection was actively carried out by all NPPs operators in China, as well as in western countries. According to the technical requirements from the regulatory bodies, many technical improvements have been implemented, including increased resistance to external flooding, improvements of emergency core cooling system and related equipment; implementation of (a) transportable back-up power supply, (b) spent fuel pool monitoring, (c) hydrogen monitoring and control, (d) emergency control center, (e) radiation environment monitoring and emergency response, and (f) external natural hazards response, etc.

Taking the improvement of flood control capability of NPP as an example, the nuclear power flooding is re-evaluated; the maximum water depth of the plant is calculated by considering conditions of design basis flood accompany with a millennium retain period rainfall. The maximum water depth of an NPP site is determined as design basis of waterproof plugging for relevant structures and buildings. Combining with the evaluation of the potential flooding of NPP, the inspection is carried out for waterproof plugging measures for galleries, doors and windows, pipe trenches and penetrations of nuclear relevant building, so that the weaknesses have been strengthened. By increasing waterproof plugging function of holes, plugging the interface between the galleries and the nuclear island, the ability of NPP resistance to external beyond design basis flooding has been further strengthened by ensuring that protection requirements of external flooding design criteria are being met.

The implementation of the above technical improvements enhances the ability of coping with beyond design basis accidents, including multiple failures, to prevent accidents similar to Fukushima Daiichi accident.

The French Nuclear Safety Authority (ASN) issued technical requirements for additional safety reviews of nuclear facilities throughout the country in accordance with the requirements of the French government, focusing on the following items: flooding, earthquakes, loss of power and loss of cooling, accident management, technical assessments and on-site verifications.

### 4.2.3 Severe accident management

In addition to improvements in operational management of NPPs, two series of guidelines have been set up both at international and national levels, to deal with severe accident mitigation and emergency.

**1. Severe accident management guidelines (SAMG)**

Nuclear power suppliers have developed guidelines for different types of power plant designs. The first was issued by Westinghouse, U. S. in 1994 and based on results of an extensive research programme on severe accidents phenomena and on the technical basis report (TBR) developed by EPRI to propose the Westinghouse Owner Group

(WOG) SAMG. This document summarizes research carried out on severe accidents management in typical PWRs in the United States. Due to its advanced technological basis and logical structure, it has been widely recognized and followed internationally.

In 2019, the IAEA issued the safety guide No. SSG-54, *Accident Management Programmes for Nuclear Power Plants*, which focuses on the requirements for the development of management procedures for preventing severe accidents and mitigating their consequences and proposes requirements for the development of guidelines for the management of severe accidents.

Since the first edition WOG SAMG has been continuously improved in its contents, guidelines, as well as application. A brand-new system PWR WOG SAMG was introduced more recently in 2016, combining the latest research results and engineering experience in the field of severe accident management for more than twenty years and making it convenient and widely applicable.

The French EDF developed the Guide d'Intervention en Accident Grave, which is mainly directed at Gen-Ⅱ PWR NPPs in France. Framatome has also developed the corresponding guidelines OSSA for severe accident management for the EPR design. The OSSA guidelines cover all the power plant stages including full power operation, low power operation, shutdown, and nuclear fuel storage in spent fuel pools.

In China, the SAMG guidelines have been implemented during the last period beginning in the year of 2000 in all NPPs operating or under construction. After the Fukushima accident, the guidelines have been expanded to include full power operation, low power operation, shutdown, and spent fuel storage facilities. Differences in design have resulted in different framework structures.

Since 2013, China Nuclear Energy Association is entrusted by China National Nuclear Safety Administration (NNSA), to organize and conduct peer reviews of SAMG at various NPPs, including Unit 1 and 2 of Tianwan NPP, LingAo Nuclear Power Plant, and Fangjiashan Nuclear Power Plant as well as Qinshan Phase Ⅲ. At the request of NNSA, NPPs carry out SAMG training incorporating a comprehensive implementation of SAMG into their regular exercise plan and special severe accident practice.

Assessments by peer experts, and special practice conducted at NPPs among other efforts have effectively improved the management capacity and emergency response level of China's NPPs, and have greatly enhanced operational safety. However, these improvements were implemented in a relatively short period of time and more efforts are needed on some issues including mutual cooperation, multi-units' accident-handling, and verification.

**2. Extensive damage mitigation guidelines (EDMG)**

Extreme damage needs to be envisaged in relation with conditions such as fires and explosions caused by terrorist attacks. These may induce extensive damage, leading to the failure of conventional accident management procedures. A possible aggression that has to be considered might be of the kind of the September 11, 2001, a terrorist destruction of the World Trade Centre towers in New York. After that event, the U. S. Nuclear Regulatory Commission (NRC) required NPPs to develop accident management strategies and guidelines for damage of this type with the objective of maintaining containment integrity and restoring reactor core and spent fuel pool cooling.

To comply with the federal regulations requirements of 10CFR50. 54, the NPPs in the United States have established extensive damage mitigation guidelines (EDMG). In this framework, it is assumed that the main control room is not accessible and that the remote shutdown station has lost the ability to control the state of the nuclear power plant, so that EOP and SAMG cease to be operational. Licensees have developed and implemented strategies to maintain or restore core cooling, containment, and spent fuel pool cooling capabilities under the circumstances associated with loss of large areas of the plant due to explosions, fire, airplane crash, etc.

The Europe has conducted "stress tests" on NPPs, to primarily assess the impact of extreme external events on nuclear facilities. These tests focus on security threats and reactor accidents caused by malicious or terrorist activities. This assessment led EDF to propose the development of FARN ("La Force d'Action Rapide Nucléaire"), under which national-level professionals and equipment form an emergency rescue team that can be quickly brought to the accident site and has the capability of simultaneously intervening on multiple units. FARN rescue needs to clarify its start-up criteria, potential tasks, configuration of emergency professionals and emergency resources, requirements for personnel training and also corresponding management procedures in NPPs. Nuclear power plants in South Africa and Spain have already established EDMG; South Korea is also researching and developing EDMG to deal with the extensive damage that might be caused by extreme external disasters.

Research institutes in China have also been, in recent years, actively engaged in the development of severe accidents mitigation guidelines. Nuclear power plants, such as Hongyanhe, Fangjiashan, Sanmen NPP Units 1&2 and Fuqing NPP Units 5&6, have already completed the development of EDMG. Other nuclear power plants are actively engaged in this development. As already indicated, one considers in this framework that the main control room and remote shutdown station lose the ability to control the state of a NPP and that EOP and SAMG cannot function. Numerous conferences organized by NNSA have been held to discuss the development and implementation of EDMG as well as its impact on nuclear power plant emergency plans. In general, China has made useful progress in the development and implementation of EDMG but the implementation experience has to be consolidated and their integration within existing emergency response systems needs to be pursued.

### 4.2.4 Insights on similar severe accidents in future

Among the three severe nuclear accidents that have occurred, the Chernobyl and Fukushima accidents have caused serious consequences.

As already indicated, the Chernobyl accident was caused by flaws in design and repeated violations of safety procedures by operators which left the reactor out of control giving rise to super prompt criticality, resulting in reactor explosion due to sharp power increase. After the accident, the nuclear industry abandoned core design concepts with positive feedback. Inherent safety of reactor was improved.

The Fukushima nuclear accident was the first NPP accident in history induced by external disaster with a high amplitude earthquake plus an accompanying tsunami. It was also the second nuclear accident in human history after the Chernobyl nuclear accident that was rated as $7^{th}$ on the INES scale.

After the Fukushima accident, China carried out an extensive safety analysis on coastal NPPs' resistance to earthquakes and tsunamis.

China is in the Eurasian continental plate, and its tectonic structure is in the inner part of the plate. The main destructive seismic activities are shallow earthquakes within the continental plate and inside the earth crust. The energies of such earthquakes are much lower than those of earthquakes in a sub-duction zone, and the deformation and displacement as a result of such earthquakes are far from those that can trigger a tsunami. In addition, China enjoys a broad continental shelf along its seaside, in which the water depth is not conducive to the accumulation of tsunami energy. Coastal conditions of China thus differ from those prevailing in Japan, both in terms of earthquake magnitude levels and high amplitude tsunamis. This is also the case for France where such extreme natural disasters have never been observed for thousands of years.

Several causes which made the Fukushima nuclear accident so severe (the flooding of the emergency pumps, the lack of catalytic hydrogen recombiners) have also been eliminated.

## 4.3 Conclusions

Environmental risks in the event of a severe accident that might occur in the future have been substantially reduced. NPPs both under operation and construction are endowed with mitigation measures, which would control the radioactive source term release and limit the impact of such accidents when they occur. These are meant to drastically reduce the area affected, so that there would be no need for permanent relocation, or emergency evacuation beyond the immediate vicinity of the plant, a limited sheltering, and no long-term restrictions in food consumption.

Comprehensive protection and mitigation measures for severe accidents contribute to a higher level of safety in Gen-Ⅲ reactors. Gen-Ⅲ PWRs are equipped with advanced large containment vessels capable of resisting external hazards such as earthquakes, tornadoes, aggressions by fires and explosions induced by humans, as well as accidental or intentional crashes by large commercial aircraft. These vessels are also able to withstand harsh internal conditions such as augmented temperature, increased pressure and radiation after accidents, and maintain their integrity, thus avoiding radioactive release to the environment.

Given the design of the nuclear power plants, the low exposure to natural disasters and the enhanced safety guidelines now implemented by NPP operators, the probability of a nuclear accident such as Chernobyl and Fukushima Daiichi, which have caused large radioactive releases, has been considerably reduced in China and France.

But an accident is precisely unpredictable. In general, any future accident would differ from previous ones. A rationale way to control the risks is to consider that accidents are possible following unexpected scenarios. Therefore mitigation features should be enhanced, to minimize potential offsite consequences, even more than is done for other human activities.

## Appendix 4 - 1 Dedicated prevention and mitigation measures for severe accidents of Gen-Ⅲ NPPs

In response to various severe accidents, Gen-Ⅲ PWR NPPs have set up a series of severe accident mitigation measures to maintain the integrity of the containment and prevent massive release of radioactive substances. Currently, the major potential causes that could threaten the integrity of containment of pressurized water reactor nuclear power plants have been identified and the corresponding countermeasures have been

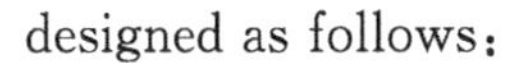



designed as follows:

(1) Direct containment heating (DCH) caused by high-pressure core melt, is avoided by a special quick safety valve generally set on the pressurizer.

(2) Hydrogen elimination systems in the form of catalytic combiners to control accumulation of this combustible gas in the containment.

(3) The reliability and redundancy of the heat-removal system is augmented to avoid the risk of overpressure in the containment. This is implemented by increasing the number of spraying system in the containment, ensuring a reliable source of water, and using a passive heat-removal system.

(4) To avoid steam explosion outside the pressure vessel, dry pit design or measures for external cooling of the reactor pressure vessel are generally used to prevent melt-through. The former eliminates the water needed for steam explosion, while the latter prevents the melted core from being released, which can fundamentally eliminate the possibility of steam explosion.

(5) In order to avoid the MCCI (molten core concrete interaction) and subsequent floor penetration, melt retention in the heap and a core melt trap may be used. The former avoids pressure vessel penetration, while the latter collects and cools the molten core after it has melted through the pressure vessel.

(6) For the failure of containment bypass, the current NPPs are designed to increase isolation reliability and design pressure of the low-pressure system. In order to reduce the amount of release, in case of a severe accident caused by the interface LOCA, it is imperative that the system required in normal operating conditions for the reactor coolant, be located within a containment-enabled building. For the bypass of SGTR (steam generator tube rupture) release, one has to take the necessary measures to prevent the SGTR steam generator from overflowing.

In particular, in the event of a severe accident with a failure of mitigation measures, with the pressure in the containment continuing to rise, the filtration and exhaust system should be able to decompress and discharge the containment in a safe and controlled manner ensuring that the pressure does not exceed the load limit. The filtering system installed on the pressure relief pipeline is designed to withhold radioactive material in the exhaust gas with filtration efficiency that reaches 99. 9% ensuring that only inert gas and a small amount of volatile substances be discharged into the environment.

## Appendix 4 - 2 Safety of inland NPP in China

Nuclear power plant can be built near sea (costal) or inland (riverside and lakeside). Except for engineering availability requirements of each plant site, nuclear power plants around the world implement unified nuclear safety assessment standards and comply with unified rules for construction and operation, so that there should be no controversy about inland nuclear power plants. At present, more than half of the nuclear power units are located in inland areas, among which, 74% of USA total 99 nuclear power units under operation are inland, 70% of French total 58 nuclear power units under operation are inland. However, Chinese mainland now still has temporary management problems for inland nuclear power.

The nuclear and radiation safety regulation and standard in China are formulated in accordance with IAEA safety standards, for maintaining, improving and enhancing NPPs to keep pace with international standards. In-land nuclear power construction based on current nuclear safety regulations in China, can meet current international highest requirements for nuclear power construction.

The definition of design basis earthquake of siting for Chinese mainland NPP has adopted strict international standards, that take into account extreme earthquakes effect and once-in-a-thousand-year seismic fortification standards. At present, the peak value of seismic acceleration at fully demonstrated inland site is less than 0. 15$g$, while the design standard for seismic design of Gen-Ⅲ PWR is 0. 3$g$, which indicates that the seismic robustness of Gen-Ⅲ PWR at these sites is significantly enhanced. In addition, inland nuclear power plants adopt the siting concept of "dry plant site", which ensures avoiding of flooding impacts. In terms of cooling water, closed circuits with wet cooling towers are used for inland plant site, so that the withdrawal of water is low and "thermal pollution" due to discharges water is prevented.

The site selection of inland nuclear power plants must strictly comply with existing nuclear safety regulations and meet the relevant requirements for gas and liquid effluents, and population distribution. In particularly, liquid effluent standards of inland NPPs are stricter than those of coastal areas, that their concentration limits are one order of magnitude lower than those of coastal areas. Nuclear power plants adopt the design of controlling the generation of radioactive waste from the sources, applies the best feasible technology to treat radioactive effluents, and realizes "Near Zero" radioactive fluid effluents discharge through comprehensive measures such as strict radioactive waste monitoring, optimized discharge management and strengthened environmental monitoring.

As mentioned in Section 4. 2. 1, Gen-Ⅲ PWR technology adopted by Chinese inland NPP has intact severe accident prevention and mitigation measures, which effectively prevent the occurrence of severe accidents and mitigate the consequences of severe accidents. Against the problem of radioactive wastewater treatment after severe accidents, nuclear industry in China has carried out extensive research. Even under extreme accidental conditions, the total maximum radioactive waste water can be generated designed for Gen-Ⅲ NPP is around 7000～10000 cubic meters. In order to prevent surrounding environmental water body from being polluted by these radioactive waste water, a series of measures are adopted in design, including, radioactive waste water storage in reactor building and nuclear auxiliary building; a number of waste liquid storage tanks and temporary waste liquid storage pools with large capacity to act as supplement or backup of safety building waste liquid storage capacity; setting of water inhibitor to prevent leakage, radioactive waste inhibitor and zeolite filters, to realize radioactive waste water sealing and isolation from surface water body under emergency conditions; reserving spaces in the site area to ensure that mobile emergency waste liquid treatment devices can be installed in time when waste liquid is produced. Through above measures, even in extreme conditions, the "storage, treatment, blockage and isolation" of radioactive waste liquid can be realized, to ensure that even in an extreme accident situation, radioactive release to the environment is under control, and environmental safety can be guaranteed.

In summary, at present, nuclear and radiation safety standards adopted for China NPPs meet the highest international safety standards, and nuclear technology implemented is in accordance with Gen-Ⅲ safety standards. The site safety of inland nuclear power plants in China is guaranteed, as long as nuclear safety regulations and standards are followed strictly, and reasonable and effective engineering measures are adopted. The impact on public and environment under normal operating conditions is within natural background levels, which is acceptable, and the environmental risk of nuclear power plant can be controlled under severe accident condition (with no permanent relocation, no need for emergency evacuation outside the immediate vicinity of the plant, limited sheltering, and no long-term restrictions in food consumption).

## Appendix 4 - 3 Emergency management after severe accidents in China

According to the *National Emergency Plan for Nuclear Accidents* (2013), in case of a severe accident, nuclear emergency organizations at all levels shall implement all or part of the following response actions in light of the nature and severity of the accident:

(1) Accident mitigation and control;

(2) Radiation monitoring and evaluation of consequences;

(3) Personnel protection from radiation;

(4) Decontamination, cleansing and medical treatment;

(5) Control of entrances and port;

(6) Market supervision and regulation;

(7) Maintenance of public order;

(8) Information reporting and dissemination;

(9) International notification and request for assistance.

Monitoring of radioactivity will be carried out on the site of the accident and in the surrounding environment (including air, land, water, atmosphere, crops, food, drinking water, etc.), and doses of radioactivity will be monitored for emergency staff and the public exposed to radiation. Furthermore, real-time meteorological, hydrological, geological, seismic and other observation (monitoring) measurement and forecast are carried out as well as accident conditions diagnosis and source investigation. Identification and monitoring of accident evolution, evaluation of radiation consequences, determination of the extent of affected areas, and provision of technical support for emergency decision-making are mandatory.

### A4 - 3.1 Three level nuclear emergency system in China

The National Nuclear Accident Emergency Coordination Committee is composed of experts from nuclear engineering, nuclear safety, radiation monitoring, radiation protection, environmental protection, transportation, medicine, meteorology, oceanography, emergency management, public communication, who provide advice and suggestions for important decisions and plans of national nuclear emergency work and for nuclear accident response work.

The government at the provincial level shall establish a provincial nuclear emergency committee, which is composed of responsible persons from relevant functional departments, relevant cities, counties and operating units of nuclear facilities, to be responsible for nuclear accident emergency preparedness and emergency treatment tasks, and to uniformly direct nuclear accident off-site emergency response actions, within its jurisdiction. The provincial nuclear emergency committee establishes an expert group to provide decision-making advice, and establish nuclear accident emergency office to undertake the daily work of provincial nuclear emergency committee. Besides, the provincial nuclear emergency front command department is established to give decision-making support.

The nuclear emergency command department of nuclear installation operators is responsible for organizing on-site nuclear emergency preparedness and treatment task, uniformly commanding nuclear emergency response action of its own, assisting off-site nuclear emergency preparedness and response task, and providing suggestions for entering off-site emergency state and taking off-site emergency protection measures.

### A4 - 3.2 Nuclear emergency monitoring system

National Nuclear Accident Emergency Command Department or National Nuclear Accident Emergency Coordination Committee shall organize national emergency forces to carry out radiation monitoring, depending upon the actual situation, which organize and coordinate national and local radiation monitoring forces to carry out radioactive monitoring in areas where is already or possibly affected by nuclear radiation (including air, land, water, atmosphere, crops, food and drinking water, etc.).

The government at the provincial level and nuclear accident emergency department of nuclear power plant should ensure radiation monitoring work after accidents, and provide support for taking emergency countermeasures and emergency protection measures for nuclear accidents.

The provincial environmental protection department has an environmental monitoring group, including land, sea, air, food, and drinking water monitoring groups. The provincial radiation environment monitoring and management station has communication, data collection and transfer devices, who is responsible for organizing and coordinating off-site emergency monitoring after nuclear accident for provincial environmental monitoring group, and for collecting and summarizing all monitoring data, analyzing the possible radiation impact of accidents on environment and public, providing monitoring data for provincial emergency evaluating center, and providing decision-making basis for provincial nuclear emergency command department.

The emergency response group of operating units coordinates and implements emergency radiation monitoring and environmental sampling to ensure that emergency radiation monitoring can be started short after accidents. The emergency response group includes monitors, samplers, people who guide and coordinate monitors and samplers and people who analyze data, sample and other information provided by monitors and samplers. There is at least one trained monitoring group per day on-site who can start emergency radiation monitoring at any time, to carry out emergency radiation monitoring. One emergency radiation monitoring group can undertake monitoring and sampling duties independently and simultaneously.

At present, the state has set up China's nuclear emergency rescue team, consisting of 6 sub rescue teams, about 320 people, which are established within the national emergency framework and rely on existing nuclear emergency forces from army and nuclear industry, who undertake the task of sudden rescue and emergency treatment task for NPP severe accident under complex conditions, effectively controls the source of nuclear accidents, searches and rescues of trapped people in time, stops the spread of the accidents with all strengths, minimizes the consequences of the hazards and supports treatment actions for nuclear facilities.

A multi-level coordinated command to be in place of the emergency command system, a unified decision-making, a multi-sector coordination after a severe accident, and a rapid deployment of emergency resources shall be present. After a severe accident, the data and information channels shall remain unobstructed, and decision-making means shall be diversified, to allow effective decision-making and support in all circumstances.

# Chapter 5 Nuclear safety and the environment

---

Recommendations

As the main goal of environmental protection is to eliminate the possibility of large radioactive releases, it is recommended that owners of nuclear facilities should:

● Test the resilience of the existing nuclear facilities to external events higher than considered in the design basis;

● Upgrade existing nuclear facilities to meet the same safety objectives as set for new facilities, as reasonably achievable;

● Implement the risk-informed defence in depth, including "beyond design basis" conditions, for all facilities;

● Perform internal and independent reviews of their safety management systems, and not exclusively rely on the reviews performed by the safety authorities.

As environment protection is a major sensitive issue for people, it is recommended that nuclear regulatory agencies should:

● Establish a transparent supervision of nuclear safety through transparent communication;

● Initiate a permanent dialog with local authorities and the public.

---

As digitalization of the nuclear industry has been progressing at a fast pace, special attention should be given to protect software and databases used at design, construction and operation stages. Nuclear Operators should identify a chief security officer (CSO), and set up, under the responsibility of the CSO, an organization dedicated to the development and implementation of a digital security policy.

# Introduction

The fundamental safety objective is to protect people and the environment from harmful effects of ionizing radiation (Ref. [42]). This is primarily achieved by controlling the radiation exposure of workers and the release of radioactive material to the environment during normal operation of NPPs and fuel cycle facilities. The smaller the nuclear facilities release of radioactive substances, the smaller their impacts on the environment. Releases are continuously monitored and controlled as shown in Chapter 2; their consequences on the environment are well below the level of natural radiation. Furthermore, they are kept as low as reasonably achievable, and records show that releases have been steadily reduced over time, to reach an asymptote at an extremely small fraction of authorizations granted by safety and environmental agencies. In France for instance, the average radiological consequences of liquid releases are in the range of $10^{-6}$ Sv/a, which is a factor of one thousand below the authorized level ($10^{-3}$ Sv/a) (Ref. [43]), which itself is 30 times less than natural radioactivity. Long-term deferred releases of radioactivity from radwaste disposed of in geological formations are also expected to lead to radiological exposition, however, much smaller than natural radioactivity as discussed in Chapter 3.

Therefore, the first section of this chapter addresses the two other objectives of nuclear safety: restricting the likelihood of nuclear accidents and mitigating the consequences of such accidents should they occur. The following three sections emphasize a few specific issues which tend to be of increasing importance nowadays, including:

- Siting in relation to safety;
- Responsibility for safety and role of the government;
- Nuclear safety and public acceptance.

Appendix 5 - 1 presents the architecture of the safety regulation system.

# 5. 1 The safety of nuclear power plants and their environmental impact

From the onset of commercial nuclear energy, safety requirements have been set up to prevent accidents and limit their consequences. Historically, the safety analysis of NPPs was based on the identification of a "design basis accident" (DBA). It was to be demonstrated to the Regulator that such accident, and any accident having a higher probability of occurrence, would result into fairly limited releases to the environment. To achieve this goal, a dual approach is taken, i. e. : ①all safeguard systems mitigating nuclear accidents less or as severe as the DBA have to be provided with adequate redundancy and diversity; and ②multiple barriers have to be set up in order to drastically limit radioactive releases to the environment. This approach, known as deterministic defence in depth has to be systematically enforced, with special attention to the independence of the safeguard systems and of the barriers (See § 5. 1. 2 Risk-informed defence in depth); in earlier designs, accidents with core melt were not considered.

Overtime, this deterministic approach was supplemented with probabilistic safety analysis, following the WASH-1400 report.

## 5. 1. 1 Severe accidents and their external consequences

When the first Generation 1 and 2 reactors (Gen-Ⅰ and -Ⅱ) were designed and built, a simple reference scenario including a DBA was considered in designing their safety systems and containment, typically the loss of coolant accident (LOCA), limited to the consideration of a "double ended guillotine break" of the primary circuit in PWRs and BWRs. However, probabilistic assessments made as early as 1975 (Ref. [44], [45]) and, unfortunately, severe accidents with core melt (such as those at the Three

Mile Island NPP (1979), the Chernobyl NPP (1986) and the Fukushima Daiichi NPP (2011) ) provided evidence that DBAs did not encompass all situations to be considered by nuclear safety. Lessons learned from these accidents resulted in back fittings of existing plants, and revisions of safety objectives.

The quantitative safety goals of NPPs have been assigned after the Three Mile Island NPP accident, such as the two "one thousandth" rule①.

The Probabilistic Safety Assessment made in NUREG-1150 concluded that:

Average probability of an individual early fatality per reactor per year:

- NRC Safety Goal: $5\times10^{-7}$
- Typical PWR: $2\times10^{-8}$

Average probability of an individual latent cancer death per reactor per year:

- NRC Safety Goal: $2\times10^{-6}$
- Typical PWR: $2\times10^{-9}$

These results could appear as quite satisfactory. However, both the Chernobyl and the Fukushima Daiichi NPP accident evidenced that nuclear safety should not only consider lethal consequences of nuclear accidents, but environmental consequences which could require evacuation and relocation of population, including people living at some distance from the plant (up to 30 km).

New safety objectives, including consideration of severe accidents are now formalized in the regulations of several countries. They have been summarized by the Western European Nuclear Regulators' Association (WENRA) as follows (Ref. [46]):

- Accidents with core melt which would lead to early or large releases **have to be practically eliminated**;
- For accidents with core melt that have not been practically eliminated, design provisions have to be taken so that only limited protective measures in area and time

① a. For normal individuals next to a NPP, the risk of immediate death due to a reactor accident should not exceed one thousandth of the total risk of immediate death caused by other accidents faced by social members; b. For the population in the vicinity of a NPP, the risk of cancer death due to the operation of the NPP should not exceed one thousandth of the total risk of cancer death caused by other causes.

are needed for the public (**no permanent relocation, no need for emergency evacuation outside the immediate vicinity of the plant, limited sheltering, no long-term restrictions in food consumption**) and that sufficient time is available to implement these measures.

Severe accidents (i. e. with core melt) can be triggered either by external events, and/or by the malfunction of engineered safety systems and hypothetically by a terrorist attack.

It is of utmost importance that operators check the resilience of their facilities to external events (flooding, earthquakes, hurricanes, etc.) more severe than the environmental conditions taken as design basis, and demonstrate that there is no "cliff edge effect" as encountered at the Fukushima Daiichi NPP. After this accident, the European Union promoted the performance of "stress tests" for all European sites, the results of which were made publicly available. These stress tests achieved two important benefits: ① they helped identifying weaknesses in the design of some facilities, and address them; and ②being made publicly available, they provide evidence to all stakeholders and the public that safety issues were thoroughly reviewed. So far, the IAEA provides no guidance on the performance of similar stress tests at all NPP; issuing a Safety Guide on this matter would help promoting their regular and systematic implementation.

With respect to engineered safety systems, and as recommended by the WENRA, it is necessary to supplement the outdated DBA approach with the consideration of other, or extended conditions, beyond DBAs. This is achieved by a comprehensive risk-informed defence in depth.

It is recommended to implement similar safety principles to existing NPPs and improve their safety levels according to the same objectives as set for new NPPs (practically eliminate large or early releases). In that respect, WENRA has provided guidelines which may be used as a reference (Ref. [47]). All French reactors are presently being upgraded as part of the large retrofit program ( "Grand Carénage" ), to meet the requirements applied to Gen-Ⅲ reactors as closely as possible. And the

Chinese government has taken provisions to enhance nuclear safety to prevent large releases of radioactive substances in case of severe accident (See Appendix 5 - 2).

### 5.1.2 Risk-informed defence in depth

It is advisable to "practically eliminate" (WENRA wording) any scenario inducing large or early releases, to drastically limit the residual risk, by such measures as increasing safety margins, adopting supplementary safety measures, and strengthening defence in depth.

Design and setup of supplementary measures should be based on the principle that nuclear safety needs be as high as reasonably achievable and ensure that such measures do not induce negative effects. To this end, various factors including the probability and the consequences of the residual risks should be comprehensively taken into consideration, and the adverse effects on response functions dedicated to normal operation, anticipated operational occurrences (AOO), DBAs and design extension conditions (DEC) should be prevented.

Risk-informed defence in depth system (RDIDS) is illustrated in Table 5 - 1. RDIDS employs engineered safety features, additional safety features and supplementary safety features:

**Table 5 - 1 Risk-informed defence in depth system**

| Levels of RDIDS | Objective | Basic measures | Conditions of NPP |
|---|---|---|---|
| Level 1 | Prevention of abnormal operation and failures | Conservative design, and high-quality construction and operation | Normal operation |
| Level 2 | Control of abnormal operation and detection of failures | Control, restriction and protection of systems and monitoring facilities | AOOs |
| Level 3 | To restrict accidents within design basis | Engineered safety features and accident response procedures | DBAs (to assume a single postulated initial event) |
| Level 4 | To control severe conditions, including prevention of severe accidents (4a) and mitigation of consequences (4b) | Additional safety features and accident management | DECs, including multi-failures (4a) and severe accidents (4b) |

Continued

| Levels of RDIDS | Objective | Basic measures | Conditions of NPP |
|---|---|---|---|
| Level 5 | Engineering rescue under extreme conditions; mitigation of consequences of radioactive releases | Supplementary safety features, guidelines for management of extensive damage condition and off-site emergency response | Residual risks |

At Level 3, engineered safety features are dedicated to DBAs, and should be implemented in accordance with the requirements of safety-grade systems and equipment.

Additional safety features dedicated to DECs are introduced at Level 4; as an example, a rapid pressure relief valve is added to the pressurize relief system to prevent high pressure core melt (HPCM), and the catastrophic early containment failure HPCM would induce. From a risk-informed perspective, additional safety features are not required to be safety grade (redundancy, qualification, etc.). As an example, the fire protection system may be considered at Level 4 to refill spent fuel pools, although this system is not safety-grade. A probabilistic assessment of the reliability of systems used in DECs may be performed to support the safety case.

At Level 5, supplementary safety features are used to prevent and mitigate the residual risk under extreme conditions. Such features include containment filtration and venting system, off-site emergency plans, mobile power sources for mitigating extensive damage consequences in NPPs, mobile pumps, water tanks, and mobile devices provided by nuclear power group and national institutions for support of emergency response in and around NPPs. In principle, supplementary safety features do not have to be safety-grade, provided their reliability is proven, and their availability is regularly checked.

Under the framework of RDIDS, the Level 4 requires including additional features dedicated to DECs in NPP design, consider their adequacy and reliability, and achieve a better balance between accident prevention and mitigation. Relevant NPP safety analysis should demonstrate that under severe accident conditions, containment can maintain its integrity and no large radioactive release to the environment would occur. Depending on the results of a plant-by-plant analysis, the installation of a containment filtration and venting systems shall be decided, if the integrity of the containment cannot be demonstrated.

At Level 5, it is assumed that the additional Level 4 of defence in depth failed, and although the objective was to practically eliminate large radioactive releases, such releases occur. It remains therefore necessary to prepare for emergency (implementation of off-site emergency preparedness to alleviate the consequences).

### 5.1.3 New safety threats

When assessing nuclear safety of operating and new plants, special considerations should be given to new threats such as cyber-attacks, and terrorism.

Cyber-attacks are not specific to NPPs, and protections should be implemented in a similar way as done for any large facility providing vital services, or having potential environmental impacts. Digitalization of the nuclear industry has progressed quite rapidly at all stages (design, construction, operation, and maintenance), and special attention should be given to protect software and databases used at any level.

Operators should assign a CSO, and set up, under the responsibility of the CSO, a dedicated organization to develop and implement a digital security policy at all level of its organization (Ref. [48]). The role of this organization should include the review of provisions taken by subcontractors in this field. A special care should be given to the instrumentation and control systems (I&C), now digitalized in modern plants, and especially to the safety of I&C. Since this system is vital for ensuring the safety of the facility, including its safe shutdown when needed, its protection requires special attention. It should not be connected to external networks, and changes and updates of this system should be subject to strict procedures, controls and re-qualification.

Terrorism is also not specific to nuclear facilities; nevertheless, its consequences would be quite serious, unfortunately. From the onset of civil nuclear industry, and under the auspices of the IAEA, considerable efforts have been drawn to prevent uncontrolled dissemination and use of nuclear material (Ref. [49]). Over the years, this system has proven to be efficient, and should be supported with determination. However, direct attacks of nuclear facilities have to be considered, as September 11 attacks demonstrated the vulnerability of our modern world to new forms of terrorism. This topic is confidential by nature, and it is essentially impossible to publicly discuss the approaches that are implemented in different countries. In principle, the same concept of defence in depth applies to this specific hazard. By design (such as earthquake resistance, robustness of a containment design to withstand a significant overpressure), nuclear facilities have the capacity to resist some external aggressions, but additional engineered features may be added to protect safety buildings, and withstand high frequencies vibrations induced by airplane crashes. What is more important, a national agency should be assigned for the responsibility to identify the safety threats to be considered; the operator shall build prevention and mitigation measures to cope with them, in cooperation with forces in charge of national security (police, army, etc.). Although the details cannot be provided, transparency calls for the concepts to be explained, and more specific information should be provided to certified members of parliament or regulatory agencies.

## 5.2 NPPs sitings

NPP siting should not only take into account power demand and plant layout, but also consider suitability of the site from a safety perspective, in all its aspects namely: ①site safety, ②environmental protection and ③emergency preparedness, as provided for by the international consensus on elementary requirements for siting of nuclear facilities. Emergency preparedness remains an important factor of a risk-oriented defence in depth.

At first, the three following aspects should be considered:

(1) The impact of external events (these events may be natural or artificially in-

duced) on the area where the site is located;

(2) The site and its environment characteristics that may affect the release of radioactive substances to people and the environment;

(3) Site factors that may affect implementation of emergency preparedness & responses.

The safety assessment of a nuclear site may be split into the following eight indicators: ①geology and earthquake characteristics; ②atmospheric dispersion; ③restricted areas and low-populated areas; ④ population distribution; ⑤ emergency plans; ⑥safeguard guidelines; ⑦hydrology; and⑧industrial, military and transportation facilities.

If the assessment using the above criteria qualifies a site as not being suitable and that its deficiencies cannot be compensated through design, site protection measures or administrative procedures, then the site must be excluded without further consideration (Ref. [50], [51]).

In order to preclude external safety hazards, NPP siting shall consider geological factors in depth to avoid geologically unstable areas such as seismic faults, areas that may be subjected to landslides, and volcanoes. It is also necessary to investigate factors such as climate and hydrology to protect NPPs from threats induced by typhoons, tsunamis, tides, floods, etc. It is also important to ensure that NPPs will always have sufficient heat sink capacity to remove the residual heat.

Moreover, issues such as the transport infrastructure to ship large equipment to the site, the local economy, and public acceptance also need to be considered in siting, although they are not safety related.

There is no difference in safety requirements for NPPs at inland sites and coastal sites, but factors that may be considered (such as typhoons, tsunamis, or dam collapse) may vary. Scenarios of extreme natural disasters facing inland NPPs may include: earthquakes and landslides, ground fissures/faults, subsidence; floods and dam break; earthquake and dam break.

With regard to the issue of how to prevent radioactive wastewater from affecting river water after accidents in the inland NPPs, abundant research has been carried out in China, resulting in the formulation of four principles for treating radioactive wastewater in the containment after accidents. The four principles for ensuring that the radioactive wastewater can be "stored", "blocked", "treated", and "isolated", are suggested to be used as supplementary safety measures for the safety design of NPPs, enhancing defence in depth of NPPs and further ensuring safety of nuclear power.

Similar research and development has been carried out in France, resulting in solutions adapted to each site and facility, and regularly reviewed. Exchanges between the French and Chinese institutes in charge of those matters should be encouraged.

After the accident at the Fukushima Daiichi NPP, development of nuclear power in China encountered some challenges, especially for inland NPPs. Due to the shortage of "good" coastal sites, some "not so good" coastal sites (especially with higher earthquake risks) are reassessed and considered as appropriate for Gen-Ⅲ NPP. Building NPPs in regions with higher seismic risks requires special attention from each party and an in-depth analysis, inclusion of safety margins, to allow for conservative decisions compatible with the required safety level.

In France, the suitability of sites is reviewed every ten years, before granting authorization to continue operation for the next ten years. For several sites, (Cadarache, Fessenheim as examples), the seismic design criteria were increased during the lifetime of the facilities; however, it could be proven that designs had sufficient margins to cope with these increased requirements without impairing safety.

## 5.3 Responsibility for safety and role of the government

### 5.3.1 The prime responsibility of the operator

There is no safety without a well identified organization responsible for ensuring safety and provided with adequate resources to discharge its duties. **"The prime responsibility for safety must rest with the person or organization responsible for facilities and activities that give rise to radiation risks"** (Ref. [52]). On a legal standpoint, this person is the nuclear licensee; through contractual arrangements, he may delegate operation and maintenance, in part or as a whole; but it is essential that the licensee-sometimes called "the owner/operator" -keeps full responsibility for controlling safety and demonstrates that he has enough resources for ensuring this role.

The complex structures of ownership of NPPs in the future, customized to accommodate financial constraints, may lead to situations where one facility has several owners, with operation delegated by contract to one of them. In such cases, a clear line of responsibility shall be established with respect to safety. The Regulator has to make sure that the owner/operator organization is clearly identified, which should be a condition to the award of a nuclear license.

To fulfill these responsibilities, nuclear operators should have adequate technological and financial resources allowing them to perform and manage nuclear safety activities. Those activities may be carried out by a considerable number of staff members within the organization, or subcontracted. To control and monitor safety related activities, it is common practice to establish safety departments or safety divisions, independent from the operational and maintenance divisions. In complex organizations (such as multiple sites; multiple units at one site), it is recommended for such safety departments or divisions to have a dual reporting line, operationally to the operational management at its level (unit, site, corporate), and functionally to the upper level of the safety organization. Furthermore, it is important that each employee working for the operator or any subcontractor be entitled to confidentially report any safety violation he might witness or be aware of, to a point of contact well identified within the organization and independent of the management line, without running any risk of sanction (whistle-blower). In very large organizations, it is also recommended to set up an independent inspection department, reporting to the top management within the organization, and performing audits of the system and self-inspections of the facilities without relying exclusively on regulatory safety authority to conduct their own regular inspections.

Whatever the safety organization is, it is of utmost importance that a strong safety culture, emphasizing the principle "safety first", be disseminated at all levels of the organization and its subcontractors, from the top management to the laypersons.

The IAEA has rightly reminded the prime responsibility of the nuclear operator to achieve nuclear safety. It would be helpful that, in connection with the World Association of Nuclear Operators, it makes basic recommendations on the best practices to be implemented by the operators to fully take in charge their duties.

### 5.3.2 The role of the government and the regulator

The role of thegovernment is to protect its people and the environment. It has to establish a legal and governmental framework for safety, including an independent regulatory body. In turn, the regulatory body grants construction and operating licenses in accordance with nuclear regulations. To check compliance with the license, the regulator performs supervision and inspections on the operator/licensee.

But these supervision and inspections do not take away the full responsibility of the operator in assuring nuclear safety, whatever the controls of the regulators are.

To implement these principles, China promulgated the *National Security Law of the People's Republic of China* on July 1, 2015, putting nuclear safety in the national security system together with political security, homeland security, military security, economic security, cultural security, social security, science & technology security, information security, ecological security and resource security etc.; and the responsibility of each party is clarified by the *Nuclear Safety Law of the People's Republic of China*, which was enacted on January 1, 2018. In France, these principles are included in the *Environmental Code* (Articles L591-1 and sq.) and consequential decrees.

## 5.4 Nuclear safety, and public understanding

Due to the complexity of nuclear power and external consequences of large accidents such as the accident at the Fukushima Daiichi NPP, the public is still haunted by "nuclear panic" and raises doubts about peaceful use of nuclear power. The not-in-my-back-yard (NIMBY) syndrome has reached an acute level for nuclear power and there is an escalating resistance and opposition to NPP projects. Public acceptance has become a bottleneck and hinder the development of nuclear power, whatever its merits with respect to cost and $CO_2$ emissions. There is a long way to go for better communicating with the public on nuclear safety.

Improving nuclear safety, to better prevent and mitigate the consequences of severe accidents is a prerequisite to further acceptance of nuclear energy. But it is also important that the public be aware of, and understands these improvements. It is an important part of a healthy nuclear development to improve public communication and raise public confidence in nuclear energy. Good public communication requires effective and transparent information, active public involvement and a permanent dialogue with local authorities and the public. Better education for the public in technical matters-starting with teachers and educators, and as soon as elementary school-should be a target of education systems.

Nuclear regulatory agencies have an important role to play in their handling of an open and transparent supervision and management of nuclear safety, and in establishing a public communication mechanism comprising "central government supervision, local authorities' leadership, enterprise implementation and public participation" . It is not the role of nuclear regulatory agencies to promote nuclear energy; but they should explain to the public how they play their role, and why they are confident that nuclear licenses can be granted. Governmental websites, as information disclosure platforms, should be improved to release relevant documents such as reports on environmental impact of nuclear projects, results of national radiation monitoring and information on project licensing. Public opinions should be widely listened to and engaged in the process of policy formulation and in the environmental evaluation of nuclear projects.

Experience indicates that openness is the basis, public participation the prerequisite, and sharing of benefits the key. If there is no benefit-sharing, it will be hard to solve the problem of NIMBY even with increasing awareness and an improved perception of risks of nuclear power.

Generally, there is no problem in public acceptance of existing NPP site expansion, probably because the local public (including local authorities) is fairly acquainted with nuclear energy and its benefits in promoting local economic and social development, while feeling no safety risk of nuclear power on the neighboring communities. Public acceptance of new NPP sites, however, may be more challenging as they have to be accepted without previous local experience.

## 5.5 Conclusions

As described in Chapter 4, the Gen-Ⅲ reactors have special prevention and mitigation measures for severe accidents, which could achieve the control of environmental risks and fully meet the requirements of nuclear safety regulations. However, it is important to explain that nuclear safety is an area of continuing learning, updating, and improvement with good experience-feedback systems.

Safety of nuclear facilities has been effectively improved through in-depth analysis of all types of incidents, internal and external, domestic and foreign, and even by borrowing best practices from other industries facing risk issues. By considering root causes of previous accidents and taking appropriate measures, potential safety risks are reduced to a large extent. After the accident at the Fukushima Daiichi NPP, the problem of public acceptance of nuclear power has become more important and even has become a bottleneck in the development of this energy. It is therefore important to explain that many additional safety measures have been implemented to reduce the risks, eliminate large radioactive release in case of a severe accident and protect the population and the environment.

## Appendix 5 - 1 Nuclear safety regulation system

The development and utilization of nuclear energy has brought new impetus to human development. At the same time, development of nuclear energy is also accompanied by safety related risks and challenges. As consequences of nuclear accidents may not be limited to one region or one country, the transnational nature of nuclear energy has to be acknowledged, and appropriate international cooperation promoted.

The nuclear safety regulation system is like a building that needs to systematically construct its foundation and structure. A typical regulation system (i. e. a nuclear and radiation safety regulatory building), as shown in Figure A5 - 1 below, consists in four cornerstones and eight pillars, also known as four crossbeams and eight pillars.

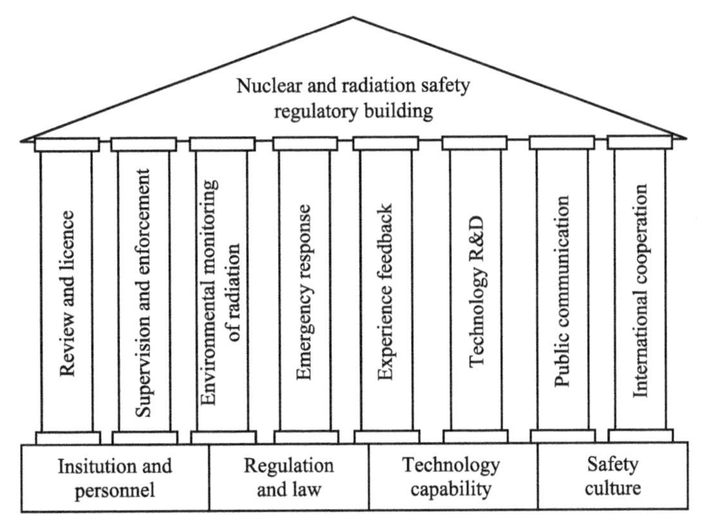


Figure A5 - 1 Schematic diagram of a nuclear and radiation safety regulatory building

The four cornerstones are institutional structure, law & regulations, technology capability and safety culture. It is generally considered that it is necessary to cement the four cornerstones as following proposals.

(1) Institution and personnel: to establish competent nuclear safety regulatory agencies independent from development departments of nuclear energy;

(2) Regulation and law: to improve top-down design of the nuclear laws and regulations based on a *Law of Atomic Energy* and/or a *Nuclear Safety Law*;

(3) Technology capability: to build platforms for independent analysis and experimental verification, information sharing, exchanges and training;

(4) Safety culture: to popularize nuclear safety culture, strengthen risk awareness, and adhere to the principle of "safety and quality first".

The eight pillars are review and license, supervision and enforcement, environmental monitoring of radiation, emergency response, experience feedback, technology R&D, public communication and international cooperation. Based on the 4 cornerstones and 8 pillars, the nuclear safety authority should build a robust and effective management system for nuclear power safety regulation.

## Appendix 5 - 2 Actions taken in China

The Chinese government promulgated the *National Security Law of the People's Republic of China* on July 1, 2015, putting nuclear safety in the national security system.

In addition, the Chinese government clearly required in the *12th Five-Year Plan for Nuclear Safety and Radioactive Pollution Prevention & Control and Vision for 2020* issued in 2012 that: new nuclear power units being or to be constructed during the period of 13th Five-Year and beyond strive to achieve the goal in design to practically eliminate the possibility of large radioactive release. In the *13th Five-Year Plan for Nuclear Safety and Radioactive Pollution Prevention & Control and the Vision for 2025* released in 2017, it is clearly stated that newly-built nuclear units will maintain international advanced level and achieve in design the goal to practically eliminate release of large amount of radioactive substances.

China National Nuclear Safety Administration released a new version of *Safety Regulations for Nuclear Power Plant Design* (HAF102 - 2016) in October 2016. HAF102 - 2016, as one of the important documents in China's nuclear safety regulation/law system, specifies binding requirements for design, specification and arrangement of structures, systems, and components important for safety of NPPs, as well as requirements for conducting comprehensive safety assessment.

HAF102 - 2016, with reference to the IAEA document *Safety of Nuclear Power Plants: Design* (SSR2/1, Rev. 1), also incorporates relevant requirements published by regulatory bodies and organizations such as the United States Nuclear Regulatory Commission (NRC) and WENRA, such as protection against malicious impact by commercial aircraft.

The 12$^{th}$ and 13$^{th}$ Five years plans state that:

(1) Under the condition of DBA and/or DEC, accidents in NPP will not result in significant release of radioactive substances.

(2) Under extreme conditions, there will be no large-scale release of radioactive substances to protect people, society, and the environment from hazards, and in particular, accident scenarios similar to the Fukushima accident which caused lasting serious pollution on the surrounding environment. The new safety goal of "practically eliminating large radioactive release" is not intended to abolish off-site emergency plan because the Fukushima nuclear accident has proved importance of the off-site emergency response. Here, the term "large amounts of radioactive release" refers to radioactive release scenarios similar to that of the Fukushima nuclear accident.

The HAF102 - 2016 regulation lays equal emphasis on the following three issues:

(1) Prevention of both internal events and external events;

(2) Prevention and mitigation of severe accidents;

(3) Deterministic and probabilistic analysis.

Major upgrades introduced in HAF102 – 2016 require to:

(1) Strengthen prevention of radiological consequences unacceptable to the public and the environment;

(2) Avoid early release and long-term pollution on the surrounding environment by taking measures for severe accident mitigation;

(3) Prevent severe accidents through NPPs design, including strengthening Level 4 of defence in depth, considering impact of external events and maintaining sufficient safety margin;

(4) Strengthen reliability of ultimate heat removal;

(5) Consolidate emergency power supply;

(6) Enhance safety of fuel storage to avoid water-uncover of fuel;

(7) Provide interfaces to facilitate uses of mobile devices where necessary;

(8) Strengthen performance of emergency response facilities.

# Chapter 6 Conclusions

The present report is a continuation of the work carried out previously by experts from the three Academies (Chinese Academy of Engineering, National Academy of Technologies of France and French Academy of Sciences).

The report published in 2017 by these academies essentially focused on recommendations about the future of nuclear energy. The present report more specifically deals with the impact of nuclear energy on the environment considering all operations from uranium mining to radioactive waste disposal. It addresses the four major environmental issues associated with nuclear power generation:

- Evaluation and control of the radioactivity released by nuclear installations under normal operation;
- Management of long-term radioactive spent fuels and radioactive waste, notably those that will be disposed of in geological repositories;
- Management of severe nuclear accidents and their radioactive releases;
- Improvement of nuclear safety as a way to limit environmental impacts and to contribute to public acceptance of nuclear energy.

On the one hand, nuclear power has many benefits, in particular that of providing an on-demand source of electrical energy and/ or heat with extremely low levels of GHG emissions. In the context of global warming, nuclear power with its near absence of GHG emissions, features a unique capacity to massively generate electricity. Furthermore, in contrast to fossil fuel plants that emit, through combustion, important quantities of air pollutants such as particles, nitric oxides, sulfur oxides, and heavy metals, NPPs do not generate air pollutants. On the positive side also, nuclear power requires a relatively limited use of land. Nuclear energy production is also flexible enough to be used for compensating a large proportion of intermittent renewable energy sources.

By summarizing these positive features, one may conclude that nuclear energy can effectively solve the problems of environmental change and constitutes one of the most appropriate sources of energy for accompanying the necessary energy transition. Without nuclear energy, the objective of GHG emission reduction seems to be difficult to attain.

On the other hand, nuclear energy may have potential adverse effects on the environment that need to be assessed, and this constitutes the focal point of this report.

It is first indicated that under normal operation, the impacts of nuclear energy on the environment are well documented and that measurements of the concentration of radionuclides in the environment are easy to do. This allows independent monitoring of such installations. The radioactivity levels of the releases are regulated in all nuclear countries according to safety rules for radiation protection. The actual releases only reach a few per cent of the authorized levels, which themselves are well below the impacts of natural radiation. This is why the report concludes that the impact of NPPs under normal operation is negligible or quite limited in terms of radioactivity.

The question of cooling water is then considered. NPPs are frequently built near the seashore and seawater is used to ensure cooling requirements. The temperature of such seawater increases slightly in heat exchanger devices before being released to the sea without any consequence.

In other cases, NPPs are sited near large rivers and the condenser is cooled either by a once through cycle (the cooling water is returned to the river) or with cooling towers. Operation of inland-sited NPPs under the first solution may face more limitations due to "thermal pollution" (increased water temperature) downstream the facility. Cooling towers drastically limit any thermal impact to the river but to the detriment of water withdrawal. The public should be better informed about measures taken to control water temperature and limit water withdrawals when one considers siting of new NPPs along large rivers.

Protection of the environment requires to be considered at each step of radwaste management;

- Isolation/confinement in packages;
- Storage, and disposal in near surface or deep geological facilities adapted to each type of radioactive waste.

Solutions rely on top level engineering technology developments, and are supported by continuous R&D on the behavior of radionuclides/toxics in engineered barriers and in the geosphere and benefit from a large international cooperation.

Monitoring is carried out during all operations from production of radwaste to its disposal in repositories, where radwaste packages are isolated from the biosphere. The background level is permanently monitored around these facilities. Feedback from their operation shows that operational releases are less than initially expected and authorized by safety and environmental authorities when the facilities were licensed.

After closure of the repositories, monitoring will continue during a test period; then safety will change from active to passive. Most radionuclides will decay in the repositories, those that might return to the biosphere will do so at a time so long that their radiotoxic impact will be negligible. While available data from analytical laboratories and underground rock laboratories are short term data, *natural analogues* provide valuable support to waste repository modeling and safety assessment: this is for example the case for natural nuclear reactors at Oklo, Gabon, that confine actinides and fission products during millions of years, or for Mediterranean archaeological glasses having resisted to erosion and leaching during thousands of years.

The main issue considered in the report pertains to environmental impacts of severe accidents that have marked the history of nuclear energy development. Issues raised by these past accidents need to be considered in a fully transparent, independent and balanced assessment. These impacts are well documented for what concerns the three severe accidents of nuclear reactors (TMI, Chernobyl, Fukushima); less well documented for the few important accidents concerning nuclear fuel cycle facilities and an effort should be made to present the feedback of these latter events. The present report indicates that on the one hand, the accidents ranked at level 7 on INES (Chernobyl and Fukushima) have had a large impact on the environment and have reduced public confidence in the nuclear energy generation system. On the other hand, the return of experience has led to important improvements in many aspects including reactor design and operational management as well as in the development of severe accident management guidelines and this has proved to be quite valuable.

The environmental risks in the event of a severe accident that might occur in the future have been substantially reduced. The NPPs, which are operating or under construction are endowed with prevention and mitigation measures that will limit the impact of such an accident if it occurs. These are meant to drastically reduce the area affected, limiting pollution and avoiding the need for a long term and large-scale evacuation of people.

One aspect that is still not settled is that of the long-term effects of low and very low dose rate exposures. There is no consensus within the scientific and nuclear communities, even though the large majority of epidemiological studies around the world converge to demonstrate that they are not harmful.

Comprehensive prevention and mitigation measures for severe accidents contribute to a higher safety level of Gen-Ⅲ reactors which are equipped with additional systems to prevent core melt, and large containment buildings capable of resisting external hazards and maintaining their integrity in case of sever accidents, thus avoiding radioactive releases to the environment.

The return of experience has led to the upgrading of existing NPPs and promoted to improve the design of new reactors together with the safety guidelines now implemented by NPP operators. It drastically reduces the probability of occurrence of a nuclear accident such as Chernobyl and Fukushima. In case of such an accident, radioactive material releases would be minimized and would not require large-scale or long term evacuations of people. It would be valuable if a global assessment by IAEA or WANO could demonstrate that a high level of upgrading has been implemented all over the world for operating NPPs.

Considering that safety management is essential to environmental protection, the report underlines that:

- The risk-oriented defence-in-depth system constitutes an improved and more complete safety methodology comprising five levels that significantly reduces the residual risks and probability of a severe accident and this in turn has an important influence on the environment.
- NPP siting should not only take into account power demand and plant layout, but should also consider suitability of the site from a safety perspective, in all its aspects namely, site safety, environmental protection and emergency preparedness, as provided for by the international consensus on elementary requirements for the siting of nuclear facilities.
- Safety Authorities play a major role in the dynamics of safety improvement and its control but the full responsibility rests on nuclear operators. Both should be engaged in a positive dialog to ensure the highest level of environmental protection.

In summary, this report is aimed at providing a balanced assessment of the impact of nuclear energy on the environment. On the one hand nuclear energy has positive effects in providing energy with a very limited level of greenhouse gas emissions without emissions of air pollutants or solid nano-or micro-particles as it is the case for energy systems using fossil fuels. This is an essential asset in the current situation where climate change induced by human activities has become one of the most difficult challenges facing humankind and where air pollution has become a major problem in many countries. On the other hand, nuclear power raises local and more global environmental issues that pertain to radwaste management and to the multiple consequences of severe accidents. Considerable efforts have been devoted to defining a sustainable management of high-level radioactive waste leading to its final disposal in geological formations. Lessons learnt from the three main severe accidents have served to improve nuclear reactor design, reduce the probability of occurrence of the release of radioactivity and make sure that the consequences to the environment remain limited if one such accident occurs.

# References


[1] International Energy Agency. World Energy Outlook 2018.

[2] http: //www. french-nuclear-safety. fr/Information/Publications/ASN-s-annual-reports/ASN-Report-on-the-state-of-nuclear-safety-and-radiation-protection-in-France-in-2017: 144, 145.

[3] NCRP. Ionizing Radiation Exposure of the Population of the United States. NCRP Report No. 160, 2009.

[4] Pan Z Q, Liu S L. Radiation Level in China. Beijing: China Atomic Energy Publishing and Media, 2010.

[5] UNEP. Radiation Effects and Sources. 2016: 29, 54.

[6] Brookins G, Douglas. Migration and retention of elements at the Oklo natural reactor. Environmental Geology, 1982, 4: 201-208. 10. 1007/BF02380513.

[7] http: //www. french-nuclear-safety. fr/Information/Publications/ASN-s-annual-reports/ASN-Report-on-the-state-of-nuclear-safety-and-radiation-protection-in-France-in-2017: 53.

[8] "Assessment on Radioactive Environmental Impact from Both Nuclear Power Chain and Coal Power Chain" group. Assessment on radioactive environmental impact from different power sources. 2017.

[9] WHO, World Health Organization. Guidelines for Drinking-Water Quality. Vol. 1. Third Edition. Geneva, Switzerland, 2004.

[10] Le tritium dans l'environnement-Synthèse des connaissances IRSN Rapport DEI 2009 – 05-https: //www. irsn. fr/FR/expertise/rapports _ expertise/Documents/environnement/IRSN _ DEI-2009 – 05 _ Tritium-environnement-synthese-connaissances. pdf.

[11] Actualisation des connaissances acquises sur le tritium dans l'environnement-IRSN-https: //www. irsn. fr/FR/expertise/rapports _ expertise/surveillance-environnement/Documents/IRSN_Rapport-Tritium - 2017_PRP-ENV-SERIS-2017 - 00004. pdf.

[12] Tritium and the environment-IRSN-August 2012-July 2017-https: //www. irsn. fr/EN/Research/publications-documentation/radionuclides-sheets/environment/Pages/Tritium-environment. aspx.

[13] 2014: Annex III: Technology-specific cost and performance parameters. In: Climate Change 2014: Mitigation of Climate Change. https: //www. ipcc. ch/site/assets/uploads/2018/02/ipcc_wg3_ar5_annex-iii. pdf page 1335.

[14] Global land outlook working paper energy and land use. https: //global-land-outlook. squarespace. com/s/Fritsche-et-al-2017-Energy-and-Land-Use-GLO-paper-corr. pdf.

[15] DoE quadrennial technology 2015. https: //www. energy. gov/sites/prod/files/2017/03/f34/qtr-2015-chapter10. pdf see note 47 page 390 and 410 for reference publications.

[16] Department of Energy. Quadrennial technology review. 2015 Sep. https: //www. energy. gov/sites/prod/files/2017/03/f34/qtr-2015-chapter10. pdf. p. 390, 410.

[17] Centrales nucléaires et environnement-Prélèvements d'eau et rejets-EDP Sciences Page 169. Jun 2014. https: //www. edp-open. org/books/edp-open-books/278-centrales-nucleaires-et-environnement-prelevements-deau-et-rejets.

[18] Life cycle water use for electricity generation: a review and harmonization of literature estimates. Published 12 March, 2013. https: //doi. org/10. 1088/1748-9326/8/1/015031.

[19] Revue Générale Nucléaire. SFEN, 08/01/2019.

[20] Synthèse de l'étude thermique du Rhône-Agence de l'eau Rhône-Méditerranée-Corse, ONEMA, Agence Régionale de Santé Rhône-Alpes, Agence de Sûreté Nucléaire, CNR and EDF. May, 2016.

[21] Le Quéré C, et al. The global carbon budget 1959-2011-Earth System Science Data Discussions 5, 2013, 2: 1107-1157.

[22] IEA Statistics © OECD/IEA 2014.

[23] IEA Energy Technology Perspectives 2017.

[24] Nuclear Energy Agency-OECD-Accelerator-Driven Systems (ADS) and Fast Reactors (FR) in Advanced Nuclear Fuel Cycles-2002.

[25] Assessment of the environmental footprint of nuclear energy systems. Comparison between closed and open fuel cycles. -Ch. Poinssot, S. Bourg, N. Ouvrier, N. Combernoux, C. Rostaing. Elsevier Energy, 2014.

[26] "Research on Key Issues of Greenhouse Gas Emission from Different Power Sources" group. Greenhouse Gas Emission from Different Power Sources in China. Beijing: China Atomic Energy Press, 2015.

[27] Les études épidémiologiques des leucémies autour des installations nucléaires chez l'enfant et le jeune adulte : revue critique-Dominique LAURIER, Marie-Odile BERNIER, Sophie JACOB, Klervi LEURAUD, Camille METZ, Éric SAMSON, Patrick LALOI - IRSN - 2008-https: //www. irsn. fr/FR/Larecherche/publications-documentation/aktis-lettre-dossiers-thematiques/RST/RST-2008/Documents/CHO3-4-Epidemio-Leuc. pdf.

[28] Spix C, Schmiedel S, Kaatsch P, et al. Case-control study on childhood cancer in the vicinity of NPPs in Germany 1980-2003. European Journal of Cancer, 2008, 44 (2): 275-284.

[29] Classification of radioactive waste. 2017, China.

[30] Fourth national report for the joint convention on the safety of spent fuel management and on the safety of radioactive waste management. 2017. China, section D. 4.

[31] Research and Development Planning Guide for High-level Radioactive Waste Geological Disposal, 2006, China.

[32] Report on the Key Consulting Project of China Academy of Engineering: Strategic Research on Radioactive Waste Management in China, 2018.

[33] Assessment of the anticipated environmental footprint of future nuclear energy systems. Evidence of the beneficial effect of extensive recycling. J. Serp, Ch.

Poinssot, S. Bourg. Energies, 2017.

[34] Report of the President's Commission on the Accident at Three Mile Island. Washington, D. C. , 1979.

[35] IAEA. INSAG-1: Summary Report on the Post-Accident Review Meeting on the Chernobyl Accident. 1986.

[36] IAEA. Environmental Consequences of the Chernobyl Accident and Their Remediation: Twenty Years of Experience. 2006.

[37] IAEA, WHO, UNDP, et al. Chernobyl's Legacy: Health, Environmental and Socio-Economic Impacts and Recommendations to the Governments of Belarus, the Russian Federation and Ukraine, Second revised version. 2006.

[38] UNSCEAR. Health effects due to radiation from the Chernobyl accident. 2011.

[39] Pan Z Q. How much impact of Chernobyl and Fukushima nuclear accidents on human health. China Nuclear Power, 2018, 11 (1): 11-14.

[40] International Atomic Energy Agency. The Fukushima Daiichi Accident-Report by the Director General. 2015.

[41] The National Diet of JAPAN, Fukushima Nuclear Accident Independent Investigation Commission. The Official Report of the Fukushima Nuclear Accident Independent Investigation Commission. 2012.

[42] IAEA. Fundamental safety principles: safety fundamentals. Chapter 2. Vienna, 2006.

[43] Electricité de France. Annual Public Information Report for the PENLY NPP. Paris, 2017.

[44] USNRC. Reactor Safety Study: An Assessment of Accident Risks in U. S. Commercial Nuclear Power Plants. Executive Summary. WASH-1400 (NUREG-75/014). Washington, 1975.

[45] USNRC. Severe Accident Risks: An Assessment for Five U. S. Nuclear Power Plants. NUREG-1150. Washington, 1990.

[46] WENRA. Safety of New NPP Designs. Study by Reactor Harmonization Working Group RHWG. March 2013.

[47] WENRA. Safety Reference Levels for Existing Reactors. Update in Reation to

Lessons Learned from Tepco Fukushima Daiichi Accident. 2014.

[48] IAEA. Computer Security at Nuclear Facilities. Nuclear Security Series No. 17. Vienna, 2011.

[49] IAEA. The International Legal Framework for Nuclear Security. International Law Series No. 4. Vienna, 2011.

[50] IAEA. Site Survey and Site Selection for Nuclear Installations. No. SSG-35. Vienna, 2015.

[51] IAEA. Managing Siting Activities for Nuclear Power Plants. No. NG-T-3. 7. Vienna, 2012.

[52] IAEA. Fundamental Safety Principles: Safety Fundamentals. Principles 1 and 2. Vienna, 2006.

# Glossary

ADS: accelerator driven system

AI: artificial intelligence

ANCCLI: Association Nationale des CLIs (National Association of CLIs)

Andra: Agence National pour la Gestion des Déchets Radioactifs (French Nuclear Waste Agency)

AOO: anticipated operational occurrences

ASN: Autorité de Sûreté Nucléaire (French Nuclear Safety Authority)

ATF: accident tolerant fuel

BAT: best available technology

BDBA: beyond design basis accident

BWR: boiling water reactor

CA: control area

CAE: Chinese Academy of Engineering

CAEA: Chinese Atomic Energy Authority

CCGT: combined cycle gas turbine

CCS: carbon capture and storage

CEA: Commissariat à l'Energie Atomique (French Atomic Energy Commission)

CEFR: China Experimental Fast Reactor

CFC: close fuel cycle

CIAE: China Institute of Atomic Energy

Cigeo: Centre Industriel de Stockage Géologique (French underground waste facility)

CLI: Commission Local d'Information (local commission delivering information to stakeholders)

CNE: Commission Nationale d'Evaluation (National Evaluation Commission of the Waste Strategy and R&D)

CNNC: China National Nuclear Corporation

CNPE: China Nuclear Power Engineering Corporation

CSA: Centre de Stockage de l'Aube (Low activity storage-Aube Department)

CSM: Centre de Stockage de la Manche (Low activity storage-Manche Department)

CSO: chief security officer

CSP: concentrating solar power

DBA: design basis accident

DCH: direct containment heating

DEC: design extension conditions

DOE: Department of Energy, USA

EDF: Electricité de France (French Utility)

EDMG: extensive damage mitigation guidelines

EPO: environmental permanent observatory

EPR: European Pressurized Water Reactor

EPRI: Electric Power Research Institute

EU: European Union

FARN: La Force d'Action Rapide Nucléaire

FR: fast neutron reactors

Gen-Ⅱ, Gen-Ⅲ, Gen-Ⅳ: refer to the second, third and fourth generations of nuclear reactors presently operated or under development (The first generation were prototypes, which are now decommissioned)

GFR: gas cooled fast reactor

GHG: greenhouse gas

GIAG: Guide d'Intervention en Accident Grave

GIF: Generation-Ⅳ International Forum

GSG: general safety guide

GWa or GWy: energy produced by one GW during one full year

HBRA: high background radiation area

HLW: high level waste

HPCM: high pressure core melt

HPR1000: advanced pressurized water reactor developed in China (also named Hualong One)

HWR: heavy water reactor

IAEA: International Atomic Energy Agency

I&C: instrumentation and control

ICRP: International Commission on Radio Protection

ICPE: Installation Classée Pour l'Environnement (facility regulated as sensitive to the environment)

IEA: International Energy Agency

IL-LLW: intermediate level-long lived waste

INES: International Nuclear Event Scale

IRSN: Institut de Radioprotection et Sûreté Nucléaire (French Technical Safety Organisation)

LBLOCA: large break loss of coolant accident

LCA: life cycle analyses

LFR: lead-cooled fast reactor

LIL-SLW: low and intermediate level-short lived waste

LLW: low level waste

LL-LLW: low level-long lived waste

LOCA: loss of coolant accident

LPCRP: *Law on Prevention and Control of Radioactive Pollution*

MCCI: molten core concrete interaction

MEE: Ministry of Ecology and Environment

MOX: mixed uranium-plutonium oxide

MSFR: fast spectrum molten salt reactor

MSR: molten salt reactor

NIMBY: not-in-my-back-yard

NISA: Nuclear and Industrial Safety Agency (Japan)

NNSA: National Nuclear Safety Administration (China)

NPCSC: National People's Congress Standing Committee

NPP: nuclear power plant

NRC: Nuclear Regulatory Commission (USA)

NRSC: Nuclear and Radiation Safety Center

OECD: Organization for Economic Co-operation and Development

OFC: open fuel cycle

OTC: once through cycle

PNGMDR: *Plan National de Gestion des Matières et Déchets Radioactifs* (Multi-annual plan for disposal of radioactive waste)

PRA: probabilistic risk assessment

PRIS: Power Reactor Information System

PV: photovoltaic

PWR: pressurized water reactor

R&D: research and development

RDIDS: risk-informed defence in depth system

SAMG: severe accident management guidelines

SBLOCA: small break loss of coolant accident

SFR: sodium-cooled fast reactor

SGTR: steam generator tube rupture

TBR: technical basis report

THMC: thermal, hydrogeological, mechanical and chemical

TMI: Three Mile Island

TTC: twice through cycle

UNEP: United Nations Environment Program

UNSCEAR: United Nations Scientific Committee on the Effects of Atomic Radiation

$UO_x$: uranium oxide

URL: underground research laboratory

VLLW: very low-level waste

VVLLW: very, very low-level waste

WANO: World Association of Nuclear Operators

WENRA: Western European Nuclear Regulators' Association

WHO: World Health Organization

WOG: Westinghouse Owner Group

# Postscript

Mid-2017, and in the wake of COP21 and COP22 committing to a significant worldwide reduction of greenhouse gas emissions, the three Academies (Chinese Academy of Engineering, National Academy of Technologies of France and French Academy of Sciences) presented a comprehensive review of the potential roles that nuclear energy could play to progressively replace fossil fuels. Its merits as a reliable and dispatchable source of electricity were outlined, and recommendations were made in the field of project management, education and training, research and technological development, to further improve the acceptance of this technology.

However, potential environmental consequences of nuclear energy are a strong concern for the public, which the Academies decided to specifically address in this second report. It considers a Life-Cycle Assessment of nuclear energy including uranium mining, reactor operation dismantling, accident consequences.

It is found that radiological consequences of nuclear operation, including the fuel cycle, are a small fraction of natural radiation, most of them from uranium mining. Recommendations are made to further reduce these consequences. A comparison of the environmental footprint of nuclear generation compared to other sources is made, which shows the merits of nuclear energy with respect to land, or material used. A specific emphasis is made to water withdrawal and consumptions which may be an issue at some river sites but can be alleviated with a proper design of the cooling systems.

Waste management is comprehensively discussed. The present waste management policies based on the latest technologies for waste confinement, including disposal of long-lived high-level radioactive waste in deep underground repositories, are analysed.

Safety analysis are summarized which show that even after thousands or millions of years, the possible additional radioactivity from the potential migration of radionuclides through the confinement barriers including the repositories themselves, the additional radioactivity will remain below about 1% of natural background radiation. Some further developments are however recommended.

Several accidents with core melt which happened at several sites in the world (Three Miles Island, Chernobyl, Fukushima) raise understandable questions from the public. Lessons learned from these accidents are presented. Safety requirements applicable to new facilities are presented. They include prevention and mitigation features to drastically reduce external consequences beyond site boundary and avoid the need for long-term evacuations of people. It is recommended that existing facilities should be upgraded to meet these requirements as closely as possible.

Improvements in safety analysis are presented, which include an extension of the traditional defence in depth to consider severe accidents with core melt both in the design and operation of reactors, to meet the "no-evacuation" goal even in the most unlikely scenarios.

Public acceptance of nuclear energy is discussed. Although it is a country specific issue, it is outlined that transparent information to a well-educated public is paramount.

This report reflects positions of the three Academies acting as independent bodies, and shall not be construed as positions of industrial actors in the NPPs field or positions of either the French or Chinese governments.